Third World Cities

What is the effect of increasing urbanisation in Third World countries? *Third World Cities* represents a thoroughly revised and updated version of a classic text which examines urbanisation in developing areas. Using case studies of cities drawn from around the world, including Bangkok, Delhi, Manila, Mexico City, Singapore and Harare, David Drakakis-Smith confronts three main questions: Is there still a Third World, does it have a common urban form and what is the relationship between urbanisation and sustainability? Starting with a discussion on whether there is still a Third World, the author goes on to analyse:

- colonial cities
- urban population growth
- the urban environment
- employment and incomes in the city
- basic needs and human rights
- planning and management.

Containing a wealth of student-friendly features including boxed case studies, discussion questions and guides for further reading, this text provides an invaluable introduction to the issues and processes of the city in the Third World.

The late **David Drakakis-Smith** was Professor of Economic Geography at the University of Liverpool from 1994–1999.

D0050495

In the same series

Routledge Introductions to Development Series

Third World Cities
Second edition

David Drakakis-Smith

London and New York

First edition published 1987 by
Methuen & Co. Ltd
Reprinted 1990, 1992, 1995, 1997
by Routledge

Second edition first published 2000 by Routledge
11 New Fetter Lane, London EC4P 4EE

Simultaneously published in the USA and Canada
by Routledge
29 West 35th Street, New York, NY 10001

Reprinted 2002

Routledge is an imprint of the Taylor & Francis Group

Typeset in Times by RefineCatch Limited, Bungay, Suffolk
Printed and bound in Great Britain by
Biddles Ltd, Guildford and King's Lynn

British Library Cataloguing in Publication Data
A catalogue record for this book is available from the British Library

Library of Congress Cataloguing in Publication Data
A catalogue record for this book is available from the Library of Congress

ISBN 0–415–19881–X (hbk)
ISBN 0–415–19882–8 (pbk)

This volume was completed in June 1999 in Keele and was David's last project before his death on 19 December 1999.

It had been in his mind for several years to update *The Third World City* but life and other projects got in the way. When he discovered that his illness was terminal David became determined that it would be done before he died. It is to his credit that he completed the task, although not entirely to his satisfaction. He said that working on the volume had helped him get through 1998, which he described as the most miserable year of his life, spent away from home and his children, Chloe and Emmanuel.

On his behalf, I would like to dedicate this book to Chloe and Emmanuel and to his students past and present, with the hope that future students will become as intrigued as David by the subject of Development.

Angela Drakakis-Smith

Contents

Plates

Figures

Case studies

● Tables

Acknowledgements

This revision was carried out while I was a staff member in the Geography Department at the University of Liverpool. I owe a great debt of gratitude to Mary Whearty who typed most of the manuscript and Sandra Mather who redrew many of the figures. I would also like to thank Bill Gould who, as Head of Department, gave me considerable time, space and encouragement to complete this work in what proved to be difficult personal circumstances.

Introduction

- Is there still a Third World?
- Third World city or Third World cities?
- Sustainable urbanisation

This book represents a thoroughly revised version of what has proved to be a popular text. But in the years between the mid-1980s, when the first edition was compiled, and the present, the world has moved on and changed considerably. It was legitimate even then to ask whether it was right to speak of the Third World city, implying that there were many common denominators; and with the collapse of the socialist Second World, the existence of the Third World itself has been frequently queried. Moreover, while development is still equated with urban-based economic growth, the costs of such investments are increasingly being questioned. In effect three questions may be asked. Is there (still) a Third World, does it have a common urban form and what is the relationship between urbanisation and sustainability?

Is there still a Third World?

The term 'Third World' had its principal origins in the search for an alternative to the polarised politics of the immediate post-war years. A third way or path between the capitalist and communist protagonists of the cold war. Thus the First World was the capitalist West, the Second World was the Soviet bloc and the Third World comprised the rest. The newly independent countries hoped to form a non-aligned movement following these principles and many of them began to adopt the collective term 'Third World' following their first meeting in Bandung, Indonesia, in 1955. Broader use of the term followed the publication of the sociologist Peter Worsley's book of the same name in 1964 but it also served to widen the meaning of the concept by incorporating into the discussion not only the common heritage of these countries (colonialism), but also a common legacy (poverty). It is this allegedly uniform condition of poverty which became the dominant image when the term 'Third World' was subsequently used, and for better or worse it is a connotation which has persisted.

But by the 1970s the commonalities between the various Third World countries were beginning to fragment quite rapidly. The OPEC countries, for example, became wealthy almost overnight through the huge price increases which their oil cartel introduced and the new high cost of oil hit other developing countries very severely. In addition, an outward flow of capital investment from European and North American multinationals was selectively channelled into a handful of developing countries, such as South Korea, Singapore and Taiwan, resulting in the rapid growth of their manufactured exports.

The consequence of these events was that differences between the developing countries widened considerably, and by the mid-1970s many commentators began to talk of more than three worlds. *Newsweek*, for example, outlined four worlds; the third encompassed those countries with significant economic potential while the fourth world consisted of the 'hardship cases'. Not to be outdone, *Time* magazine suggested five worlds: here the third comprised the oil producers, the fourth constituted other resource-rich states, while the fifth contained the 'basket-cases'.

Differences between the nations of the Third World have continued to grow and contrasts in Gross National Product or standards of living have widened, with many African states, in particular, experiencing real deterioration in economic and social conditions. Some academics, such as Wolfgang Sachs (1990), feel that this diversity is inevitable and, indeed, resent the attempts of the last 40 years to convert the diverse economics and cultures of the world into western look-alikes. Much of this apparent convergence of development goals, lifestyles and cultures is, of course, very patchy and shallow over large parts of the Third World. Relatively few cities have been as successful as Singapore or Hong Kong in replicating the skyline of Manhattan. Satellite links, fax machines and mobile telephones mean nothing to Bangladeshi squatters or to Ethiopian villagers walking for weeks to obtain food. Globalisation has been a very bitty process (see Case Study A).

Despite such variations, some are firm believers in there being one world not three, four or more. The seventeenth-century Czech

Case Study A

Globalisation and our shrinking world

Globalisation is often assumed to mean an intensification of linkages between economically advanced nations and the rest of the world, with the former becoming increasingly dominant in the course of this process. There are, however, several facets to globalisation. First, there is the argument that national political boundaries have been transcended by the activities of regional political entities, such as the EU or APEC, and/or multinational corporations. Second, there is the driving force of economic integration itself as transnational operations draw capital, physical and human resources from around the world into a global network of production and marketing. Third, there is the convergence of local and national cultures into a lumpy but increasingly homogeneous hybrid that is dominated, like all the other aspects, by the West.

All of these features depend on the notion of a shrinking world, the assumption that time and space have been collapsed by the acceleration of developments in transport and communications, particularly in the second half of the twentieth century. A hundred years ago it took a traveller a month to reach Cape Town from Britain and two months to reach Hong Kong, now the journey to each can virtually be done overnight. Add to this the almost instantaneous communication of satellite television and electronic mail and it would appear that there is nowhere in the world that is not immediately contactable from everywhere else.

But this conclusion would be wrong. It is only in selected places and for specific people that the world has shrunk to this extent. For those who have little to offer the global marketplace, exclusion rather than inclusion is far more common. Thus the most convenient way to fly from Kano in Nigeria to Harare in Zimbabwe is via London. Inclusion or access to the shrinking world is, moreover, often set within a framework not of increasing choice but of increasing control by a small, influential group of gatekeepers who may be media barons or politicians. Many national newspapers or television stations may use sophisticated production technology, but their content is heavily censored or subject to manipulation or massage. When local controls are not possible, international linkages are denied, as with Singapore's attempted exclusion of the Star satellite system.

Far from creating one world through more rapid and intensified interlinkages, the process of inclusion or access to, and exclusion from, the means by which the world has allegedly shrunk has, in the view of many, widened the gap between the haves and have-nots, whether at the household or national level. In contemporary Vietnam, for example, re-entry into the world economy since the late 1980s has occurred at a bewildering range of different rates for various social groups. Some can relax with a cold Pepsi or Heineken while they watch live satellite soccer from the new Chinese superleague. Others no more than a few kilometres away still take as long to walk to market as their great grandparents did, and are forced to pull their children out of school because they cannot afford the US 50 cents monthly fee. For these and many other people the world is still changing relatively slowly. As one Vietnamese sociologist expressed it 'with power and access to the market you can change your life, without that you can't get rich!'

philosopher Comenius stated that 'we are all citizens of one world, we are all of the same blood . . . let us have but one end in view, the welfare of humanity'. These sentiments have been echoed many times over the years, not least by eminent politicians from developing countries, such as Indira Ghandi and Julius Nyerere. Such Utopianism, others would claim, flies in the face of widening inequality on a global scale. The ratio between the wealth of the top 20 per cent of the world's nations and the bottom 20 per cent has steadily widened from 30:1 in 1960 to 60:1 in the 1990s. We may all be in the same boat, but few are on the top deck and fewer still are steering; most are doing all the hard work, receive optimistic information about the destination of the vessel, but never seem to get anywhere (Figure 1). Many observers, including the major development agencies, prefer a dualistic alternative to summarise world contrasts. The most influential of such classifications has been the North–South terminology employed in the Brandt Report of 1981 (Figure 2). This seemingly straightforward divide has been criticised by many, however, for perpetuating the image of a wealthy, modern set of donor nations on the one hand, and a poor, backward set of

Figure 1 *All in the same boat?*

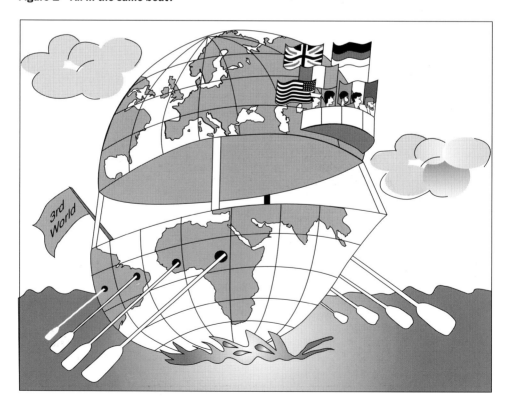

Figure 2 *North and South*

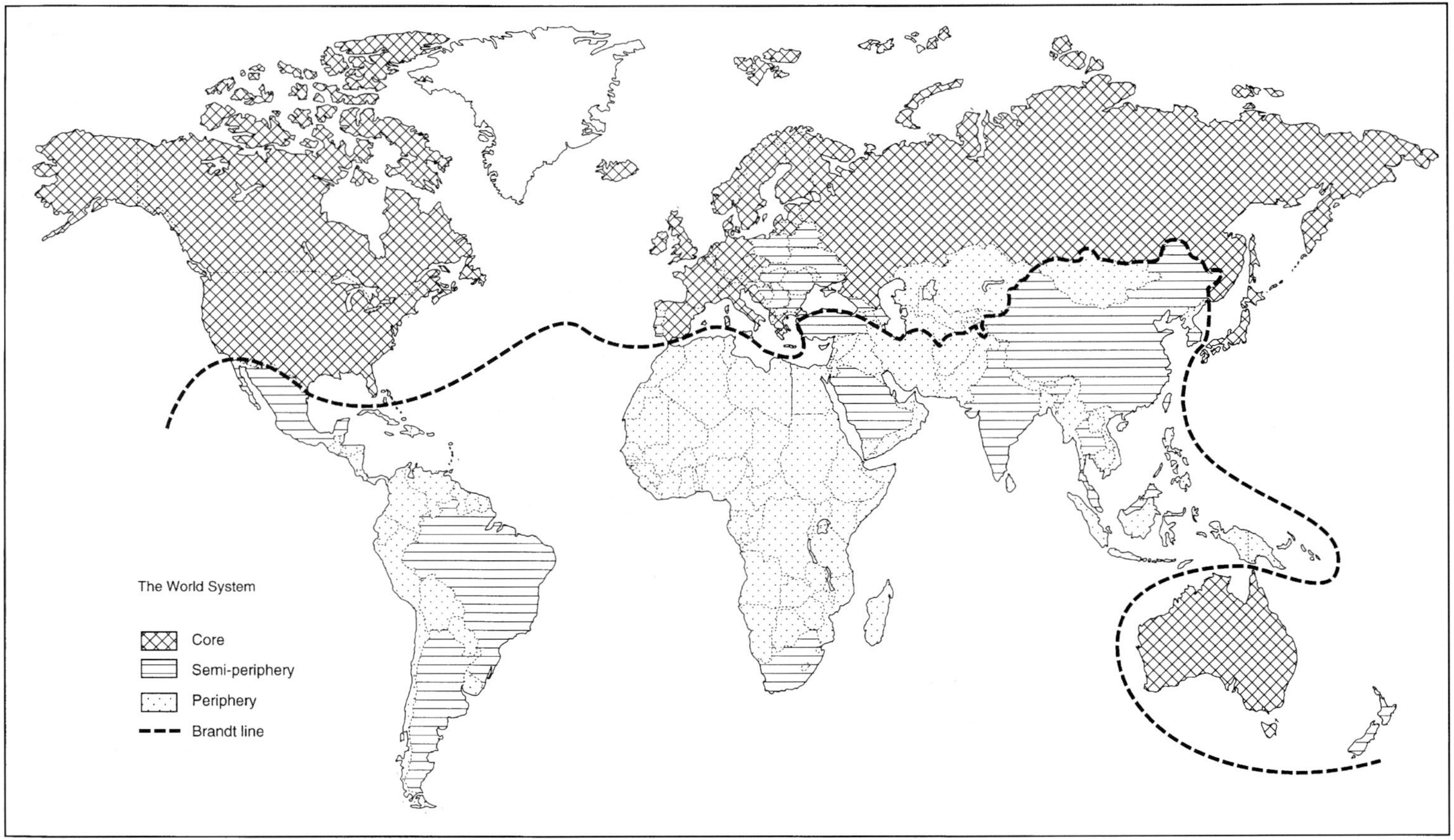

The World System

⊠ Core
▥ Semi-periphery
⬚ Periphery
- - - Brandt line

aid-receiving nations on the other. Such myths have been difficult to dispel.

So where do all these alternative terms and concepts leave us in the late 1990s, particularly with the virtual disappearance of the Second World? Jim Norwine has argued that diversity need not destroy a common identity and sometimes strengthens it. He uses the example of the tropical rainforest which has a huge range of flora and fauna but which, nevertheless, constitutes 'one organic whole' far greater than the sum of its parts. So it is with the Third World which is united by its colonial past and its continued subordinate role in the world economy. Most people in most developing countries still live in grinding poverty and their numbers are growing. Fundamental economic and political relations at the global level sustain this situation. In this sense the concept of the Third World is an 'extremely useful figurement of the human imagination . . . the Third World exists whatever we choose to call it. The more difficult question is how to understand it' (Norwine and Gonzalez, 1988) and change it for the benefit of the poor.

Third World city or Third World cities?

Cities in the Third World also exhibit considerable diversity, from the nature of their built environment to features such as their ethnic composition or economic structure. Much of this contrast will be the result of quite different histories, both of the individual cities and their respective countries. Some cities, such as Cairo, were large and powerful long before European colonialism began, others are essentially the creation of the present century, such as Harare or Nairobi. Local ethnic and cultural circumstances have also helped to maintain diversity in contemporary urban development, so that despite the growing similarity of many downtown areas to their western counterparts, cities in Sri Lanka or Myanmar offer quite a different urban experience from those in Algeria or Colombia.

Third World cities also vary considerably according to their place in the urban hierarchy within their own countries. Capital cities, whether large or small, usually have a common array of global links, even if only through diplomatic or retail channels. The lifestyle pursued in such cities is often much more westernised in its nature than that in smaller settlements, particularly those in economic backwaters or poorer regions where the division between rural and urban lifestyles is less marked. Nevertheless, even the remotest town usually abounds with advertisements for western cigarettes or soft drinks.

So despite the myriad differences between cities throughout the Third World, common processes can be identified which have shaped and continue to shape their nature. Amongst these commonalities are the legacies of colonialism and the sheer pace of demographic change, each of which is discussed within this book. However, the factor which impacts most of all on Third World cities is their linkage to the global economy. While such links are many and varied, the fact that they are structured around a global capitalism in which developing countries have particular roles to play, means that there are many common political and economic forces at work in shaping Third World cities. Such forces would include the impetus to create a built environment, particularly in capital cities, which is designed to impress foreign investors as much as to meet the basic needs of its inhabitants; they might also encompass the policy wishes of major aid donors, for example in encouraging the adoption of structural adjustment programmes in order to overcome debt problems which has resulted in substantial reduction of funds for urban basic needs programmes (Case Study B). While these and other processes may not operate to the same extent, if at all, in each city, they are sufficiently influential to create in the developing countries as a whole a degree of convergence in which common processes are seen to produce similar consequences – high-rise city centres, burgeoning squatter settlements, flourishing informal sectors and the like.

And yet, we must not forget that for each city these processes are located in specific historical, geographical and cultural settings which always give local variations to what may seem like a familiar problem. Ethnic conflict may fragment squatter settlements and prevent collaborative self-help schemes from succeeding; religious constraints may affect the incorporation of women into the urban economy. Third World cities are, therefore, shaped by a diverse combination of local, national and global processes and it is the purpose of a book such as this to illustrate how these processes operate in the real world. In this task it is always helpful to have more specific foci around which the discussion can occur – a middle way between broad theory and local detail that brings both together in a process of mutual information. The chapters in the volume thus present a series of major themes relating to Third World cities and discuss them in this light.

Sustainable urbanisation

It may be helpful in identifying such themes and their degree of integration to consider briefly the sustainability of the kind of urbanisation that is currently occurring over much of the developing world. While the sustainability of the development process as a whole has been a fashionable and justifiable concern since the mid-1980s, urbanisation has been seen simply as one of the villains of the situation, contributing to the world's environmental problems by its insatiable demand for more resources, and by spreading its wastes over the surrounding areas. But with more than half of the world's population already living in towns and cities, and with most development strategies being firmly focused on urban-based economic growth, it is appropriate to consider the sustainability of the contemporary urbanisation process itself.

In order to do this we must have a set of goals whereby we can evaluate the urbanisation process and its sustainability. At this point we must make a clear distinction between 'sustained growth' which is often an urban-focused development process driven by economic goals, and 'sustainable urbanisation' which, as with the broader objectives of sustainability, aims to meet present needs 'without compromising the ability of future operations to meet their own needs' – as the Brundtland Commission noted.

What are the needs of the present? While recognising that cities do play a central role in the development of most states and that economic growth is a necessary and justifiable development objective, urban sustainability must also satisfy the following requirements:

- equity, social justice and human rights
- basic human needs, such as shelter and health care
- social and ethnic self-determination
- environmental awareness and integrity
- awareness of linkages across both space and time, i.e. not seeking gain at the expense of someone elsewhere in the world or of the generations to come.

Sustainability also emphasises the interlinked nature of the individual components of rapid urbanisation (Figure 3). This means that as far as possible urban management policies need to be aware of the implications of changes in one part of the system for the remainder. It is, for example, futile to invest in improved health care facilities if those treated continue to live in squalid squatter huts

Figure 3 *The components of sustainable urbanisation*

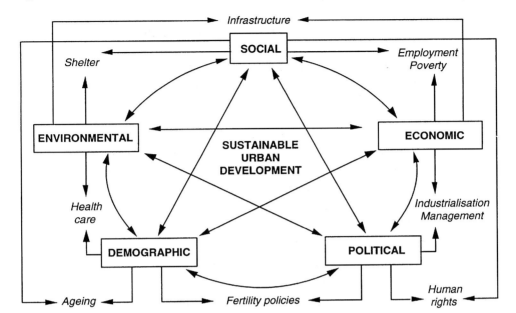

Main component, for example, Social
Issue, for example, *Shelter*

without access to clean water or adequate sanitary facilities. It is also true that economic growth must bring returns to the individual citizen in the form of jobs and income, as much as it brings returns to the firm or the state. Recognising this inter-relationship of economic, social and cultural processes is essential to the management of sustainable cities.

In each of the themes pursued in this book, this interpretation of urban sustainability must be borne in mind. The merging of global and local development processes can produce diverse consequences, some of which have already produced urban situations which are as undesirable as they are unsustainable. Only by identifying such problems, their global and local causes and the nature of their integration, can we hope to change the situation for the better for all the residents of the urban Third World. Expanding and improving our awareness of these needs is the principal objective of this book.

Key ideas

1. Although a growing diversity of interests has characterised the developing countries over the last quarter of the twentieth century, common developmental experiences can still be identified.
2. The concept of 'The Third World' is a highly debated topic.
3. Cities in the Third World exhibit both differences and similarities often linked respectively to local and global development processes.
4. Urban sustainability is a growing problem within the Third World.

Discussion questions

* Has growing diversity destroyed the validity of the Third World as a concept?
* Has the world shrunk into one uniform economic and social system?
* Is there such a thing as 'the Third World city'?
* What is sustainable urbanisation and why is it so important?

References and further reading

Allen, J. (1995) 'Global Worlds', in J. Allen and D. Massey (eds) *Geographical Worlds*, OUP, Oxford: 105–142.

Allen, J. and Hamnett, D. (1995) *A Shrinking World*, OUP, Oxford.

Drakakis-Smith, D. (1995) 'Third World cities: sustainable urban development', *Urban Studies*, 32(4–5): 659–677.

Drakakis-Smith, D. (1996) 'The nature of Third World cities', in I. Douglas, R. Huggett and M. Robinson (eds) *An Encyclopaedia of Geography*, Routledge, London.

Norwine, J. and Gonzalez, A. (1988) *The Third World: States of Mind and Being*, Unwin Hyman, London.

Potter, R., Binns, A.J., Elliot, J. and Smith, D. (1999) *Development Geographies*, Addison-Wesley, Longman, London.

Sachs, W. (1990) *The Development Dictionary*, Zed Books, London.

The dimensions of urban growth in the Third World

- Questions of definition
- Levels of urban population
- Rates and scales of growth
- Extended metropolitan areas
- Urban growth and economic development
- Conclusion

In 1990, for the first time, the World Bank recorded as many people living in urban settlements as in the countryside, with the global growth rate of urban populations rising at 4.5 per cent each year. As Figure 1.1 reveals, however, the principal component of this global growth comprises the low income countries. While the urbanisation process refers to much more than simple population growth and involves the examination of associated economic, social and political change, the starting point for such analysis is the demographic change which has brought so many people together in such a relatively short time; it is to this topic that the present chapter addresses itself.

Questions of definition

There is undoubtedly great diversity in the nature of urban growth in the Third World, as the Introduction pointed out. Even capital cities can range from small agglomerations of less than 20,000 in Pacific island states to the massive 20 million of Mexico City, while average annual urban growth rates during the 1980s and 1990s have oscillated between those in excess of 10 per cent, in Tanzania for example, and those of less than 2 per cent, such as in Sri Lanka. This diversity also extends to definitions of what is 'urban' or what constitutes a 'city'. Each country tends to adopt its own criteria and these can vary enormously. In Peru, for example, an urban settlement is one which exceeds a hundred occupied dwellings; in India the figure is around 5,000 people. The definition clearly affects the overall percentage of

Figure 1.1 *Urban population as a percentage of total population, 1965–1995*

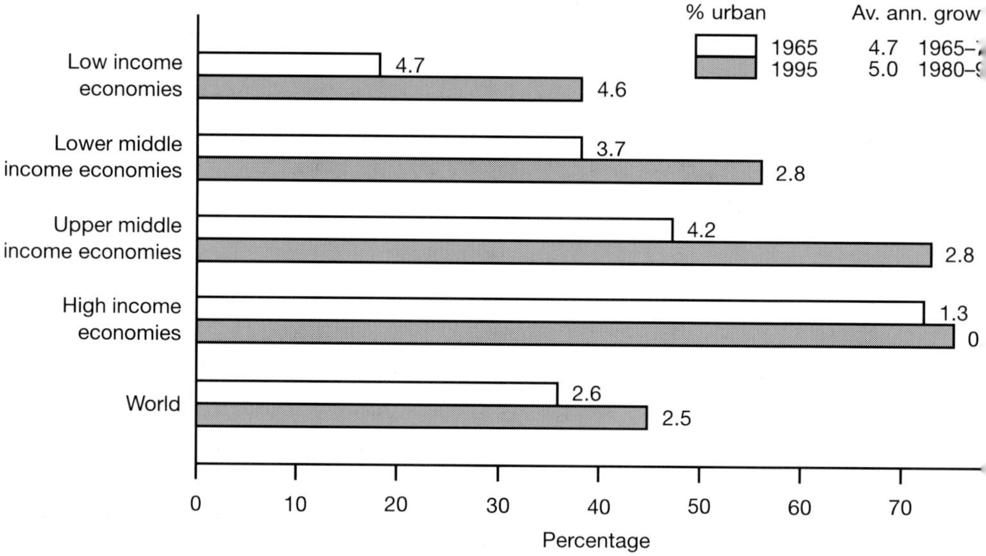

Source: World Bank, *World Development Report* (1997).

the population officially classified as 'urban'; in Peru it currently stands at 71 per cent, in India it is only 26 per cent and yet India contains many more cities with more than one million inhabitants.

Such variations make meaningful international comparisons very difficult, more so when urban totals are artificially affected by boundary issues. In China, for example, many cities incorporate extensive rural areas within their boundaries in order to ensure control over vital urban resource needs, such as water or energy. As a result, many official census returns often incorporate rural populations who have little direct contact with the urban core. Shanghai, for example, incorporates large areas of agricultural land within its designated area of over 6,000 square kilometres (UNCHS, 1996). These situations are further compounded by inaccurate registration systems, together with frequent changes to boundaries and official definitions of what is 'urban'. In Vietnam, as a consequence of such problems, the official urban percentage of the total population showed a decline from 19.8 per cent to 19.4 per cent between 1989 and 1995, when in fact all other indicators revealed rapid urban growth, particularly in the largest cities.

In an effort to overcome these statistical variations, the various international agencies that collate information have at times attempted to standardise their data. The United Nations, for example, has recognised settlements of over 20,000 as 'urban', of more than 100,000 as 'cities' and of more than 5 million as 'big cities', but these standardisations have not become common usage. Other agencies, such as the World Bank, simply accept national definitions, and perhaps this is the appropriate approach. If such data reflect what each country considers to be historically, culturally and politically 'urban', then so be it. Why should some arbitrary, often westernised statistical base line be officially imposed?

Levels of urban population

Figure 1.2 reveals very clearly that there are wide differences in the proportion of national populations that can be classified as urban. In very general terms some three-quarters of the population of developed countries are urban, compared to only one-third in developing countries; in the least developed countries only around 20 per cent of the population is urbanised. Of course, such broad figures conceal considerable variations within the Third World and many parts of Latin America are as urbanised as Europe, with individual cities such as São Paulo being amongst the most populous in the world.

In contrast, most of Africa is far less urbanised, containing many countries where more than 70–80 per cent of the population still live in rural areas. Variations to this general picture may be found in the north of the continent where a much longer history of urbanisation in countries fringing the Mediterranean has given rise to higher contemporary proportions. Asia appears to fall between Latin America and Africa in terms of level of urbanisation and also to be a little more uniform in its urban character, but a distinct split has emerged over the last decade or so between the South Asian bloc centred on India and the Pacific Asian bloc, including China. In the latter, urban population levels are much higher than in the former, a phenomenon which has been associated with industrialisation. But whatever the national proportion, Asian countries tend to be characterised by both very large cities and by dense rural populations, often in the areas around these cities. The contrasts between the urban and rural ways of life are consequently at their most intense in many parts of Asia.

There are, therefore, considerable variations within and between the different regions and countries of the Third World with respect to the level of urban population. Moreover, these contrasts have varied

Figure 1.2 Urban population as a percentage of total population, 1998

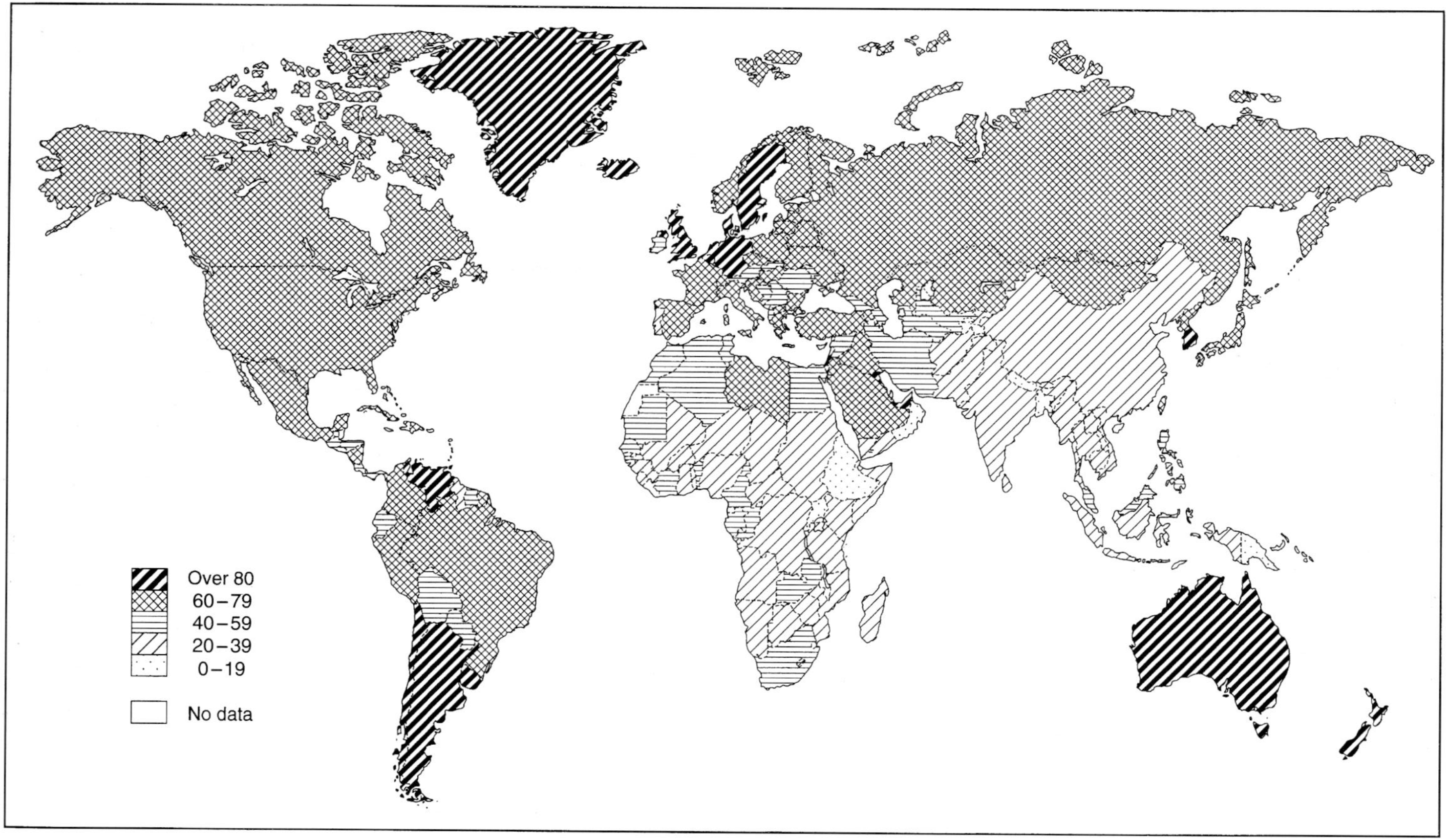

Over 80
60–79
40–59
20–39
0–19
No data

Table 1.1 *The regional distribution of the world's population in 'million-cities' and the location of the world's largest 100 cities, 1800, 1950 and 1990*

	Urban population (%)		Population in million-cities (%)		Number of the world's 100 largest cities		
	1950	*1990*	*1950*	*1990*	*1800*	*1950*	*1990*
Africa	4.5	8.8	1.8	7.5	4	3	7
Eastern Africa	0.5	1.7	—	0.8	—	—	—
Middle Africa	0.5	1.0	—	0.8	—	—	—
Northern Africa	1.8	2.8	1.8	3.2	3	2	5
Southern Africa	0.8	0.9	—	0.8	0	1	0
Western Africa	0.9	2.6	—	2.0	1	0	1
Americas	23.7	23.0	30.1	27.8	3	26	27
Caribbean	0.8	0.9	0.6	0.8	1	1	0
Central America	2.0	3.3	1.6	2.7	1	1	3
Northern America	14.4	9.2	21.2	13.1	0	18	13
South America	6.5	9.7	6.7	11.1	1	6	11
Asia	32.0	44.5	28.6	45.6	64	33	44
Eastern Asia	15.2	19.7	17.6	22.2	29	18	21
South-eastern Asia	3.7	5.8	3.4	5.6	5	5	8
South-central Asia	11.2	14.8	7.0	14.6	24	9	13
Western Asia	1.8	4.1	0.6	3.3	6	1	2
Europe	38.8	22.8	38.0	17.9	29	36	20
Eastern Europe	11.8	9.3	7.7	6.3	2	7	4
Northern Europe	7.7	3.4	9.0	2.1	6	6	2
Southern Europe	6.5	4.0	6.7	3.2	12	8	6
Western Europe	12.8	6.2	14.6	6.2	9	15	8
Oceania	1.1	0.8	1.6	1.3	0	2	2

Source: UNCHS (1996).

enormously throughout the last two centuries, with the 'interruption' of the rapid urbanisation process during the industrialisation of Western Europe now being overtaken once more by Asia and with the extreme pace and scale of Latin American urban growth becoming ever more prominent (Table 1.1). But if the broad contrast made earlier between the contemporary developing and developed world holds true, why should we be so concerned with urban sustainability in the Third World when nations are much more likely to be heavily urbanised in the West? The answer, of course, lies not in the level of urbanisation but in its sheer scale and rate of growth.

Rates and scale of growth

Between 1950 and 1975 the urban population of the Third World grew by 400 million and by 2000 will have increased by a further billion. Little wonder that in terms of absolute numbers, there are now twice as many urbanites in the Third World as there are in developed countries (Figure 1.3). Perhaps more importantly, the rate of increase of these respective totals is such that in low-income countries urban growth rates exceed those of the developed world by more than five times. Figure 1.4, therefore, indicates quite a different spatial pattern compared to Figure 1.2 – almost a reverse image and more illustrative of the countries and regions where rapid urban growth poses severe problems. To put the comparison into historical perspective, the urban growth in Europe (including Russia) throughout the whole of the nineteenth century amounted to some 45 million people, a total

Figure 1.3 *World urban population composition, 1950–2000*

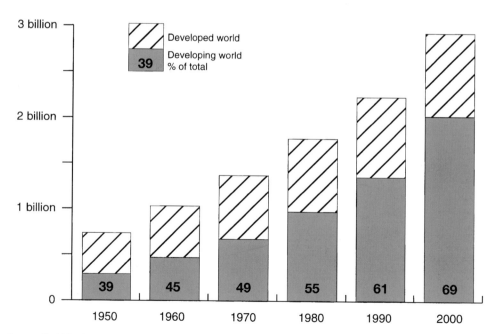

Source: World Bank, *World Development Report* (1997); UNCHS (1996).

Figure 1.4 *Annual average growth rate of urban population, 1980–1998*

Over 8.0
6.0–7.9
4.0–5.9
2.0–3.9
0.0–1.9

No data

exceeded by China alone during the 1980s. The principal causes of this unprecedented growth will be discussed in Chapter 3.

One apparently undeniable feature of contemporary urban population growth is the way in which the largest cities appear to have been growing at the most rapid rates. A phenomenon which has given rise to the concept of urban primacy – the demographic, economic, social and political dominance of one city over all others within an urban system. It is undoubtedly true that by the end of the twentieth century the world's 20 most populous cities will have switched from a Euro-American focus to a clear Third World bias within a mere 20 years. It is also true that in some developing countries primacy is very marked (Case Study B). In Vietnam, for example, it is variously estimated that between one-third and one-half of all urban dwellers live in either Ho Chi Minh City or Hanoi, giving the country one of the proportionately smallest but most concentrated urban populations in the world. However, we must not overemphasise this phenomenon. Most urban dwellers in the Third World do not live in mega-cities. Urban primacy is not, therefore, exclusive to developing countries; it has been, and continues to be, a feature of many European countries such as Greece, Austria, Portugal and Ireland. Moreover, there is extensive evidence from Latin America that the growth rates of the largest cities have slowed down considerably as problems of sustainability, such as poverty, become more serious.

In the past there has been a preoccupation with the urban primacy in the Third World and this led to the emergence of theories in which such features were said to be illustrative of serious developmental problems. Affected countries were alleged to be 'overurbanised' or possessed of 'abnormal' urban hierarchies very different from the more balanced rank-size hierarchies found in European and North American urban systems. Such comparisons revealed the weakness of these concepts – they were and are heavily based on Eurocentric 'norms'. Is there really an optimal size of city or urban system against which Asian or African settlements may be judged to be abnormal – surely not. This is, of course, not to minimise the real problems that rapid urban growth has posed for many Third World cities, but it does indicate that they must be assessed in their own regional and national contexts rather than measured against an inappropriate and often mythical European or North American standard. Nothing illustrates more clearly the need to understand the local context and process of urbanisation before the problems posed to developers and urban planners can be fully appreciated.

Case Study B

Bangkok: the most primate capital city in Asia

Until the recent economic downturn, Thailand has had the fastest growing economy in Pacific Asia, expanding at an average of some 8 per cent per year from the mid-1950s to the mid-1990s. Investment has flowed into the country from around the region, notably from Taiwan and Hong Kong, to tap its cheap labour and exploit its lax industrial and environmental legislation. Normally economic growth and urbanisation are closely related, but this is not always the case and in Thailand the situation is dramatically different. According to official statistics around 80 per cent of Thailand's population is still rural, and, what is more, almost 60 per cent of the urban population reside in one city – the capital, Bangkok.

Most observers agree that in reality Thailand probably has a higher proportion living in its towns and cities, with enumeration being made difficult by the extent of seasonal or circular migration. However, a more complete statistical count would probably only serve to emphasise the domination of Bangkok, which is the target for most migrants. Two-thirds of the migrants from the poverty-stricken north-east, for example, prefer to move the 600–900 kilometres to the capital rather than move to a regional centre.

Table B.1   *Thailand: inter-regional poverty*

	Percentage of total population						
Region	*1975–6*	*1980–1*	*1985–6*	*1988–9*	*1990*	*1992*	*1994*
North	33.2	21.5	25.5	23.2	16.6	13.6	8.5
North-east	44.9	35.9	48.2	37.5	28.3	22.3	15.7
Central	13.0	13.6	15.6	16.0	12.9	6.0	5.2
South	30.7	20.4	27.2	21.5	17.6	11.8	11.7
BMR	7.8	3.9	3.5	3.4	2.8	1.3	0.8
City core	6.9	3.7	3.1	3.3	2.0	1.1	0.5
Whole kingdom	30.0	23.0	29.5	23.7	18.0	13.1	9.6

Source: Dixon (1999).

The reason for this dominance is not difficult to discern, since most economic investment is focused on the Bangkok Metropolitan Area (BMA) and more than three-quarters of the gross industrial output comes from there (Wongsuphasawat, 1997). Not surprisingly incomes are correspondingly higher (Table B.1) and act as a magnet for both temporary and permanent migrants. In addition the capital has a disproportionately higher share of almost all other economic opportunities and social services. For example, almost half of the physicians live in the BMA and more than three-quarters of the pharmacists; two-thirds of all the banking and insurance services and three-quarters of all passenger cars are registered there; for secondary education it is almost a necessity to move to Bangkok, which produces over 80 per cent of the country's graduates. Not surprisingly, perhaps, the capital is many times larger than other cities in Thailand, none of which yet approach a fraction of its size (Table B.2).

Table B.2 *Population ratio between Bangkok and the second city*

	Bangkok	BMA	BMA region
1900	11		
1910	12		
1920	13		
1930	14		
1940	15		
1950	23		
1960	25		
1970	33		
1980		36	
1985		27	39
1990		28	42
1993		30	47

Source: Dixon (1999).

The Thai government is aware of the problems that such urban and economic concentration creates – the huge pressure on housing, infrastructural and other facilities in Bangkok and the growing dissatisfaction with life in the regions. Often this regional resentment takes on political dimensions, fuelled as it is by the fact that many of Thailand's regions have large non-Thai populations. All this is not helped by the fact that the benefits of economic growth have not been distributed evenly in social terms (as well as spatial terms). The income gap between rich and poor in both rural and urban areas has widened and the richest 10 per cent of the population receive almost 40 per cent of the country's income, whereas the bottom 20 per cent receive less than 5 per cent (Dixon, 1999).

In a *laissez-faire* market economy such as Thailand's there is little scope for direct government intervention to address the regional, sectoral and social imbalance. Some commercial redevelopments have spread into regions such as the north-east to take advantage of cheap village labour, but their impact is limited and localised. Most government schemes relate to the development of the areas immediately adjacent to the BMA or to the coastal zone to the south of nearby Chonburi. Here, as in Bangkok itself, economic expansion is virtually uncontrolled and every year brings another outrage as workers are killed in factory fires or migrant hotels collapse, while prime agricultural land is continually stripped of its top soil to enable the rural brickworks surrounding the capital to meet the insatiable desire for new building materials. Little wonder that for other countries in the region Bangkok has become synonymous with all that is worst in rampant uncontrolled urbanisation. The economy may be growing, but the nation is paying a heavy price.

Extended metropolitan areas

In recent years some observers have suggested that the nature of the urbanisation process has been changing. There are various terms used to describe the settlements associated with this process, such as mega-cities, desakota (Indonesian for city-village) and extended metropolitan regions (EMR). These agglomerations are not the same as 'world cities', a term which describes the key command and control points of the global economy, such as New York, London or Tokyo (UNCHS, 1996). EMRs are also the product of the globalisation of the world economy but refer more specifically to a pattern of urbanisation and city structure that is claimed to be fundamentally different from earlier types of urbanisation (McGee and Robinson, 1995).

Essentially EMRs represent a fusion of urban and regional development in which the distinction between what is urban and rural has become blurred as cities expand along corridors of communication, by-passing or surrounding small towns and villages which subsequently experience *in situ* changes in function and occupation. In Delhi, such villages have even maintained administrative independence, and this illustrates one of the major problems of EMRs – the lack of co-ordinated urban management of such rapidly growing agglomerations. Another major feature of EMRs is the fact that they tend to have multiple nuclei rather than one central Business District. Major functions decentralise into special centres for business, production, tourism, finance, entertainment and the like.

Terry McGee (1995) suggests that there are three broad forces that have acted together to influence the emergence of EMRs – structural change, globalisation and the transactional revolution, each of which also affects the sustainability of EMRs. Structural change refers to the transformation of economic activity and employment from agriculture towards industry and services (Figure 1.5). In some countries this change has been very rapid indeed, funded by multinational investment. As noted earlier, however, such trends have been uneven and the shift from agriculture to secondary activities has been slow and limited in many countries. Nevertheless, urbanisation has continued apace even in the poorest countries, and extended metropolitan regions have emerged despite their lack of economic strength.

Globalisation is closely linked to the structural changes described

Figure 1.5 *Structural changes in Gross Domestic Product*

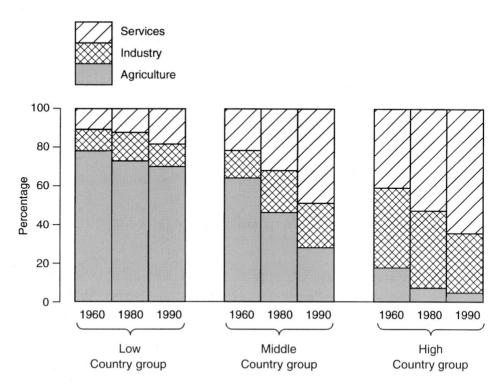

above (see Case Study A), with the internationalisation of investment, markets and even labour forces occurring in patchy fashion, and cities at all levels competing with one another in the quality and efficiency of their built environment in order to attract investment. Thus the largest cities acquire their world trade centres, telecommunications complexes, fashionable retail chains and the like.

In many ways these changes are dependent on the third of McGee's influencing factors – the shrinking of distance and time by numerous changes in transport and communications. Not all transactional changes are occurring at the same rate nor do they affect all parts of the world. Information, capital and decision-making may now be transmitted electronically, but only between those plugged into the new global networks. Moreover, people and goods still have to be physically moved, while much of the exchange of increasingly complex information must still be undertaken face to face. The impact of such varied changes on urban form is often contradictory. New transportation developments have increased the rate of immigration and also widened commuter belts in the Third World; improved electronic communications have focused much of this population movement onto the bigger cities. On the other hand, cheap personalised developments such as motor cycles, mobile telephones and PC's have meant that within cities individuals have become less place-dependent, producing an 'unfocused concentration' of people in the EMRs.

While there are common processes, EMRs are at widely different stages of development and are strongly influenced by local economic, cultural and environmental controls. Nevertheless, even the most recently emerging states have their large cities exhibiting signs of regionalisation, largely as a result of international links, as Hanoi (Figure 1.6) clearly indicates.

Urban growth and economic development

In the past many of the attempts to 'explain' rapid urban growth have relied heavily upon the apparently clear-cut links between the level of Gross National Product (GNP) per capita and the urban proportion of total population. As Figure 1.7 indicates, these two phenomena have a direct graphic correlation which is all too often interpreted as a causal relationship between urbanisation and economic development. But nothing could be further from the truth because GNP per capita is an indicator of economic *growth* rather than *development* (growth with equity), and urban population levels do not reflect the complexity of the urbanisation process. Moreover, which way does the causal relationship operate? Is economic growth a cause or a consequence of urban growth? The graph can give no indication of such influences, which will vary from city to city, and simply indicates that a general relationship exists.

Once again, therefore, aggregated data are shown to be rather ambiguous, and devoid of explanatory function. This can be emphasised by comparing Figure 1.7 with Figure 1.8 which graphs *rates* of urban and economic growth. Such a graph still lacks explanatory power but, as the previous sections have indicated, its data comprise more useful indicators of the variation in urban pressures within the Third World. Thus it is evident from the graph that, if anything, an inverse relationship exists between the two sets of statistics and that some of the most rapid urban growth is occurring in the poorest countries – those least able to cope with the resultant pressure on jobs, services and the like. The quality of life for most people in Kinshasa, Dhaka or Rangoon is desperately low, with squatter or slum housing being the norm rather than the exception,

Figure 1.6 *Hanoi's extending metropolitan region*

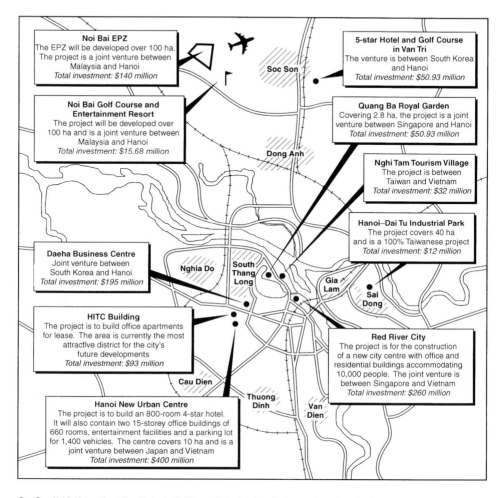

Noi Bai EPZ
The EPZ will be developed over 100 ha.
The project is a joint venture between
Malaysia and Hanoi
Total investment: $140 million

**5-star Hotel and Golf Course
in Van Tri**
The venture is between South Korea
and Hanoi
Total investment: $50.93 million

**Noi Bai Golf Course and
Entertainment Resort**
The project will be developed over
100 ha and is a joint venture between
Malaysia and Hanoi
Total investment: $15.68 million

Quang Ba Royal Garden
Covering 2.8 ha, the project is a joint
venture between Singapore and Hanoi
Total investment: $50.93 million

Nghi Tam Tourism Village
The project is between
Taiwan and Vietnam
Total investment: $32 million

Hanoi–Dai Tu Industrial Park
The project covers 40 ha
and is a 100% Taiwanese project
Total investment: $12 million

Daeha Business Centre
Joint venture between
South Korea and Hanoi
Total investment: $195 million

HITC Building
The project is to build office apartments
for lease. The area is currently the most
attractive district for the city's
future developments
Total investment: $93 million

Red River City
The project is for the construction
of a new city centre with office and
residential buildings accommodating
10,000 people. The joint venture is
between Singapore and Vietnam
Total investment: $260 million

Hanoi New Urban Centre
The project is to build an 800-room 4-star hotel.
It will also contain two 15-storey office buildings of
660 rooms, entertainment facilities and a parking lot
for 1,400 vehicles. The centre covers 10 ha and is a
joint venture between Japan and Vietnam
Total investment: $400 million

Soc Son • Dong Anh • Nghia Do • South Thang Long • Gia Lam • Sai Dong • Cau Dien • Thuong Dinh • Van Dien

Soc Son: Noi Bai International Airport is located in this area. Malaysia is investing in two projects there: A golf course and an EPZ

Dong Anh: At the moment the area is one of the main vegetable supplying sources for residents in Hanoi

Nghia Do: Only five minutes from the West Lake, a property development area

South Thang Long: Attracts a lot of foreign investment. The infrastructure is good, with the water supply system aided by the Finnish government

Gia Lam Area: Called the 'Daewoo area', it will soon become a satellite city of Hanoi

Cau Dien: Recently approved project of traditional cultural tourism villages

Thuong Dinh: Local industrial area

Van Dien: An industrial park with infrastructure not yet developed. No foreign investment at present

limited job prospects, and mortality/morbidity figures which are worse than those in many rural areas.

This pessimistic picture poses the important question of why urban population growth should be so high in countries that offer so little in their cities. The reason is, of course, that prospects are even worse in

Figure 1.7 *Levels of urbanisation and GNP per capita*

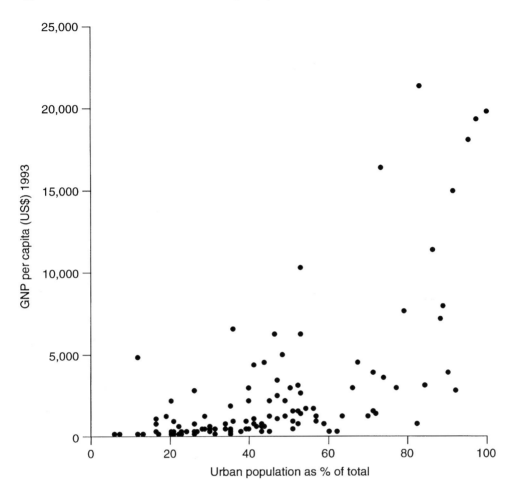

much of the countryside and it is the perceived advantages of the city that draw migrants to it. However, we must not over simplify the process of rural–urban migration, which is a complex and changing phenomenon that is dealt with in more detail in Chapter 3. It is important to note, at this stage, that many national governments also perceive cities in the same light as individual migrants – as important centres in which economic growth is generated. As much of this growth is perceived to be fuelled by foreign investment, so many governments in the Third World have, as an integral part of their national development strategy, focused their own investment priorities on the built environments of their largest cities in an attempt to impress and attract the multinational firms that they hope will generate economic growth. This results in misplaced priorities for

Figure 1.8 *Rates of urban population and GNP per capita growth*

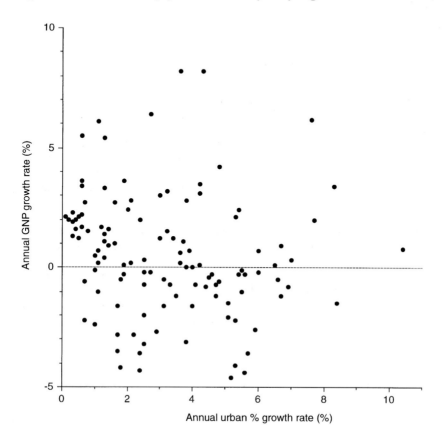

luxury hotels, conference centres or show-piece stadia, while basic needs such as health care, shelter or proper sewerage systems, particularly in low-income areas, are restricted to minimal and erratic investment schemes that bode ill for overall urban sustainability (Case Study C).

Conclusion

What this overview has indicated more than anything else is the diverse nature of global and also Third World urbanisation. In particular, it is clear that models used to formalise, represent and analyse the past, present and future urban growth patterns of Europe and North America have little relevance for the examination of Third World cities. Not only are their functions and structure quite different, but they exist within different parts of the world economic system.

Case Study C

Urban mismanagement in Manila

The Philippines were claimed for Spain by Magellan in 1521 and by the 1570s the role of Manila as the primary economic and political focus for this multiple-island state was firmly established. The subsequent colonial transfer to the United States in 1899 and to independence some 50 years later have only served to emphasise this pre-eminence, particularly since the creation of the National Capital Region (or Metro Manila) in the mid-1970s. At present Metro Manila contains almost ten million people, more than one-quarter of the country's urban population, and is ten times larger than Davao City, the second ranking urban centre.

Since independence the Philippine government has been in the hands of traditional, powerful land-owning families, but the Marcos regime in particular focused development priorities on the capital – not only to confirm its national power but also to enhance the international reputation of the 'first family'. Imelda Marcos, wife of President Ferdinand Marcos, became governor of Metro Manila and instigated a World Bank funded analysis of urban planning strategies. Not surprisingly these recommended major infrastructural investments to create a capital city in the image of the western cities that the Marcoses were so keen to imitate and impress.

Plate C.1 *Empty evidence of the Marcoses' ambitions: the International Conference Center, Manila*

In the late 1970s enormous sums were spent on large-scale projects, often linked to specific international events. Many of these projects were concentrated on a large area of reclaimed land south of the Pasig River intended as a national trade and cultural focus. The International Convention Center is typical of these projects (Plate C.1); built at a cost of US$150 million to house an International Monetary Fund conference. At the same time some US$360 million was loaned by government agencies for the construction of 14 new hotels to house the delegates to the conference and the visitors to subsequent events such as the Miss World Pageant and the Ali–Frazier fight (the Thriller in Manila!).

To put these projects and expenditure into perspective, in 1976 only US$13 million was allocated by the government for low-cost housing construction. Moreover, some 160,000 squatters also lost their homes as part of a 'beautification' programme designed to make poverty less visible to the international tourists. Some of the squatters were resettled in sites up to 30 kilometres from the city, with no shelter, water or sewage facilities. There was much resistance to these high-handed and unsympathetic actions by the government, particularly through the Ugnayan, a city-wide squatter representative organisation. Eventually, the middle class too was alienated by the activities of the Marcos regime and it was overthrown in 1986.

The legacy of the Marcos years was a weak economy, limited government reserves and an unsympathetic attitude towards urban welfare programmes. Until recently this situation had changed very little so that half of the country's population was estimated to be in poverty in the early 1990s, with the growth of the urban poor outstripping that of the rural poor. Almost one-third of the urban population of the country lives in slum or squatter settlements and the situation is improving only slowly (Cohen, 1990).

Plate C.2 *Abandoned because of ghosts: Film Center, Manila*

An important factor in this slow rate of change is the fact that the Philippines has had to adopt the structural adjustment policies recommended by the World Bank for countries in debt and seeking further loans. This has meant a cut-back in government funds for basic needs and retrenchment in government jobs, both of which have hit Metro Manila very hard. The corollary of this approach is the encouragement given to the informal sector to step into the gap left by the retreat of the state, encouraging and assisting the poor to help meet their own needs. In a country such as the Philippines with a long tradition of self-help, this has resulted in a mushrooming of NGOs and community-based organisations. However, such groups have made relatively limited progress in improving the life of the urban poor. Moreover, as is typical of many NGOs, they are often more beholden to their funding sources than to their clients, the poor, and are frequently accused of simply being agents of privatisation of what should be state responsibilities. Meanwhile, out on the reclamation at San Jose, apart from another Miss Universe contest in 1994, the slumber of the guards at the Convention Center goes undisturbed, the loose roof rattles on the Cultural Center and the Manila Film Center has been abandoned because of ghosts (Plate C.2). Not withstanding such experiences, President Fidel Ramos plans to concentrate future government energies on what are called 'flagship projects' such as new expressways around Manila and a new international airport for the capital. *Plus ça change . . .!*

None of this is to deny that much of the urban change which has occurred in the Third World over the last two decades has drawn it closer to the forces that shape western cities. The adoption of western building design, urban planning and consumption values have, therefore, begun to produce a truly global style of capital city in which central business districts are dominated by the same skyscrapers, inhabited by the same international banks, whose employees wear similar clothes and visit McDonald's or Pizza Hut for lunch. In this respect Kuala Lumpur and Suva are little different from Rome or Los Angeles. However, the incorporation of Third World nations and cities into the world system, whether capitalist or socialist, has also been historically specific and is strongly influenced by local forces and circumstances. This crucial factor must be borne in mind as we try to tease from the complexities of the real world some of the common denominators in this process. In this context there is no better or more appropriate place to start than the historical background to contemporary urbanisation in the Third World.

Key ideas

1. Problems exist in defining what is 'urban' and in distinguishing between urban population growth and urbanisation.
2. Considerable spatial variations exist in the urban proportion of total population within and between developed and developing countries.

3. The level and rate of urban growth are different phenomena and do not show similar global patterns. They each pose problems for urban sustainability.
4. A western bias has characterised past analyses of urban growth. Some suggest that new non-western patterns of urbanisation are occurring in the developing world.

Discussion questions

* Compare Figures 1.2 and 1.4 and account for the major differences.
* What is the nature of the relationship between economic and urban growth?
* What is urban primacy and what relevance does it have for the study of urbanisation in the developing world?

* What are EMRs and how are they different from urbanisation in the West?
* For any one country assemble data on the level of urban population and its rate of growth. How and why have these changed over the second half of the twentieth century?

References and further reading

Cohen, M. (1990) 'A menu for malnutrition', *Far Eastern Economic Review*, 12 July: 38–39.

Dixon, C. (1999) *The Thai Economy*, Routledge, London.

Gilbert, A. and Gugler, J. (1991) *Cities, Poverty and Development*, OUP, Oxford.

Hardoy, J. and Satterthwaite, D. (1986) 'Urban change in the Third World', *Habitat International*, 10(3): 33–52.

McGee, T.G. (1995) 'Macrofitting the urban regions of ASEAN', in T.G. McGee and I. Robinson (eds) *The Mega-Urban Regions of Southeast Asia*, UBC Press, Vancouver: 3–26.

McGee, T.G. and Robinson, I. (eds) (1995) *The Mega-Urban Regions of Southeast Asia*, UBC Press, Vancouver.

UNCHS (1996) *An Urbanizing World*, Habitat, OUP, Oxford.

Wongsuphasawat, L. (1997) 'The extended Bangkok metropolitan region and uneven industrial development in Thailand', in C. Dixon and D. Drakakis-Smith (eds) *Uneven Development in Southeast Asia*, Ashgate, Aldershot: 196–220.

2 An historical perspective

- Definitions and framework for investigation
- Mercantile colonialism
- Industrial colonialism
- Late colonialism
- Early independence

Definitions and framework for investigation

The terms 'colonialism' and 'imperialism' tend to be used interchangeably and indiscriminately in most descriptions of pre-independent development. Moreover, discussions on the colonial city are often restricted to the nineteenth and twentieth centuries. At its simplest, colonialism refers to the assumption of political control by one group over another and has existed as long as there have been identifiable political states. Imperialism essentially encompasses the political economy of capitalism, following its emergence from Europe in the sixteenth century, which some believe has followed a clear series of cycles of growth and recession (Taylor, 1985). Both began well before the nineteenth century, and the interlinkage of the various forms of colonialism and imperialism provide a useful framework within which to examine colonial urbanisation, the nature of the colonial city and the legacies bequeathed to the contemporary Third World (Dixon and Heffernan, 1991)

Table 2.1 presents a sequence of chronological phases in the colonial urbanisation of Asia and Africa; Latin America was somewhat different as its formal colonial phase ended by the mid-nineteenth century. Not every country experienced the same processes in the sequence suggested here, so that Table 2.1 simply provides a generalised summary of events. However, the important point is to realise that the nature of colonialism and imperialism changed considerably over time, initiating new processes which

Table 2.1 *Phases of imperialism and colonial urbanisation, Asia and Africa*

....

Imperialism	*Colonial urbanisation*
1500 Commodity capital: plunder of precious metals, trade in high value commodities	1500 Mercantile colonialism: limited colonial presence in pre-existing settlements, trade usually in national products of local regions
1800 Transitional phase: reduced European interest in overseas investment, greater profits from European industrialisation	
1850 Money capital: profits from Industrial Revolution re-invested in new sources of raw materials, food and markets	1850 Industrial colonialism: the scramble for territory and the creation of new urban hierarchies and more extensive European settlement
	1920 Late colonialism: development of colonialism in depth, intensification of European urban settlement, increased scale of public building, introduction of town planning, build up of indigenous urban population and segregationalism
	1950 Neo-colonialism: rapid growth of indigenous urban populations, limited employment opportunities and heavy pressure on basic needs provision
1970 Finance capital: multinational capital invested in manufacturing in selected developing countries	

....

had far-reaching repercussions for contemporary Third World cities.

Not only did colonial urbanisation vary through time, it also varied enormously over geographical space according to the complex mix of countries and cultures involved, together with the particular motivation for colonial expansion into a particular region. It would certainly be incorrect to assume that pre-colonial Asia, Africa or Latin America comprised a blank slate of backward societies on which a common pattern of colonialism was to be inscribed. A few areas did consist of more traditional societies structured around subsistence agriculture, but most contained substantial urban settlements of varying degrees of economic and political complexity.

Indeed, some cities, such as Beijing or Delhi, were considerably more sophisticated than the capitals of the European countries that began to venture overseas in the sixteenth century.

Nor were the European powers themselves of a uniform character. Although commonly seeking economic profit, firstly through trade and later from raw material production, their methods of exploration, exploitation and administration varied considerably according to cultural or political traits (for example, Britain preferred indirect rule, whereas France incorporated her colonies directly into a centralised political system), and to the nature of the agents or organisations involved. Many individuals simply sought plunder, companies sought trading commodities, national governments sought spheres of influence. Not surprisingly, colonial settlements too varied tremendously. Although we regard large ports as archetypical, there were many other types of colonial settlement, such as recreational hill stations, railway towns or military strongpoints. Even the built environment could vary too, according to prevailing architectural fashion in the metropolitan power at the time of construction. Thus Spanish and Portuguese colonial cities often had a Mediterranean air about them, with extensive promenading areas or baroque churches. In contrast, Dutch and British architecture tended to be more stolid and, certainly in the nineteenth century, reproduced the neoclassical public buildings evident in almost every British industrial town. However, colonial urbanisation began long before the nineteenth century and this chapter will use the chronology outlined in Table 2.1 to examine how and why such variations occurred. We now require a conceptual framework to guide this analysis.

In this context, it is important to realise that colonial cities were not just a physical reflection of economic and cultural change but were agents of societal transformation in their own right. Indeed, writers such as A.J. Christopher (1988) discuss the nature of colonialism largely through its settlement patterns. For Christopher there were two main types of colonial settlements – those found in largely European settled areas and those developed in the colonialisation of populous pre-existing societies. In this latter context many different types of urban form were produced dependent on factors such as the nature of the indigenous culture, the extent to which it was urbanised, the scale of the colonial presence, the integration of planning practices, and so on. But the purpose of studying the colonial cities is not to develop typologies but to understand the global and regional development processes of which they were a part, and the role they played in this.

There have been relatively few attempts to provide an analytical framework for colonial urbanisation. Most are far too empirical, setting out long lists of features against which to measure individual cities. These resemble typologies more than concepts or theories. Perhaps the most commonly used approach is still that of Anthony King (1976), who suggested that culture, technology and political power were the most useful channels through which the processes of colonial urbanisation could be analysed. Unfortunately King's ideas were related primarily to the industrial and late colonial periods; nevertheless, we will use and reassess his model at an appropriate point in our overview.

However, King (1990) does remind us to set our examination of the colonial city into a variety of spatial contexts (Figure 2.1) as its role and function often varied according to the scale of analysis. He suggests that we should look at the colonial city

- in its own right – its built environment, form and functions
- for its role in the immediate geographical region – its impact on production, transport system, settlement patterns and labour movements
- for its role in the colonial territory overall via administration, exploitation, in the hierarchy of cities, indigenous–colonial conflicts and regional imbalance
- for its links with the metropolitan power via trade, transfer of

Figure 2.1 *Spatial contexts of the colonial city*

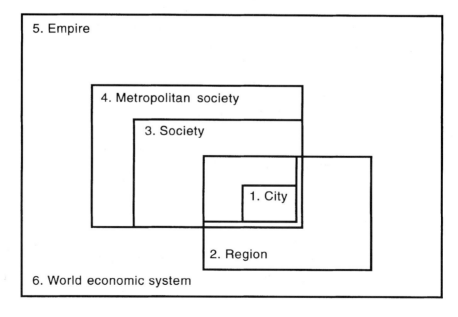

planning ideas, culture and as an intermediary in the balance of administrative power over the colony

- as part of Empire – a defensive bastion such as Singapore, a transport node such as Aden, an imperial gateway such as Bombay
- as part of an emerging world system facilitating the expansion of capitalism and the transfer of resources

The first three of these relate to the internal role of the city within the colonial territory; the second three relate more to its external linkages.

Mercantile colonialism

Initial forays from Europe were often made by individual adventurers in order to obtain inherently valuable objects such as gold and silver. Subsequently the search shifted to commodities that were valued within the European trading system, such as spices, silks or sugar. Frequently these were obtained not by trade but by simple plunder. For the most part, however, the commodities drawn into the European system were the natural products of their country of origin. This was important because their local collection was left in the hands of existing traders whose networks remained intact and were simply incorporated into the new European trading systems.

Overall, therefore, there was little need for an extensive European presence on foreign soil, although the nature of mercantile colonialism varied enormously due to the wide range of countries involved and the lengthy period it lasted. In Latin America the earliest contacts were almost entirely destructive of indigenous population and property – whole cities in the Aztec and Inca empires. Elsewhere, particularly in South-east Asia, contact was based on the trade of valuable commodities, often of Chinese origin since no direct European contact was permitted with that empire. The limited European settlement was not only due to efficient local trading networks but also to the fact that mercantile colonialism was based on private companies rather than state enterprises. Permanent company representatives were relatively few in number and tended to be confined to spatially limited concession areas within existing indigenous cities.

This situation did not remain static, however, and as profits expanded, so the Europeans began to seek a more extensive presence in order to oversee the collection of the traded commodities. Eventually these warehouses would be protected with troops, first indigenous and later

European, whose military technology was such as to ensure superiority despite inferior numbers. Later on, the demand for commodities of reliable quality led to the gradual incursion of European companies into the production process itself, extending their influence well outside the trading concession. Particularly prominent in this process were the various East India companies that operated out of several European states, such as England, France, Holland and Sweden.

In morphological terms, however, Europeans were usually confined to very small areas of the cities in which they resided. Most of these cities were already organised into ethnic/occupational quarters prior to the European arrival so that one more group made relatively little difference. For example, when the Portuguese assumed control over Malacca in 1511 this merely resulted in a bishop replacing the imam, a governor superseding the sultan, and a fort and church being constructed above the town (McGee, 1967). Trading in the port continued uninterrupted in the distinct Chinese, Javanese, Arab and Indian quarters. In other cities, where only concessionary rights were obtained, the European impact was even less marked.

There was little planning of the European areas in the contemporary sense of the word. In the early years most Europeans sought simply to reproduce familiar urban forms. For example, the Dutch in Batavia (Jakarta) constructed tall, closely packed houses along its canals, as in the Netherlands; however, diseases resultant from the malarial water and the lack of fresh air soon forced an adaptation of the long, low villas of the Javanese aristocracy. Similar trends elsewhere produced a distinctive series of colonial vernacular architectural styles that were European in function and facilities but indigenous in design and materials. In broader terms, the mercantile colonial period had little impact on systems of cities; no new hierarchies were created (unlike in later periods) and settlements of purely colonial origin, such as Lima, Manila or Cape Town were exceptions.

By the end of the eighteenth century, European interest in overseas ventures began to moderate, for several reasons, and a transitional period in the nature of colonial development emerges. First, the extensive European wars of the Napoleonic period retained within Europe many of the adventurers and some of the venture capital on which mercantile exploration depended. Second, the shift into production per se rather than trade began to inflate the cost of colonial activities beyond the capacity of the companies involved. Several of the largest and most famous, such as the Dutch, French,

Swedish and British East India Companies, went into liquidation around the turn of the century and their operations were taken over by the respective state governments.

But the principal characteristic of the transitional period was the interest of European investors in the greater profits to be made from the Industrial Revolution rather than mercantile colonialism. In this context it is not surprising that Francis Light, Stamford Raffles and Charles Elliot found the British government less than enthusiastic about the development of their 'acquisition' of the islands of Penang, Singapore and Hong Kong respectively. Palmerston's famous dismissal of Hong Kong as 'a barren island with hardly a house upon it' encapsulated the feeling of the investors, both private and public, of the period.

Industrial colonialism

Such sentiments did not persist for very long. The rapid growth of the Industrial Revolution in Europe led to increasing demands for both raw materials and food for the burgeoning urban workforce. By the third quarter of the nineteenth century European capital was once again being invested overseas, but on this occasion the principal agent of organisation was the state. The accumulation of raw materials and food required more than trading toe-holds; it depended on the acquisition of territory and the organisation of production in order to keep costs as low as possible. Much of the food was produced in colonies where European settlement was more extensive. In the British case this was Australia, Canada and South Africa. Elsewhere, physical constraints restricted European settlement to the management and administration of raw material production. Often this was no longer of indigenous crops but of introduced commodities, such as rubber in Malaysia and Indo-China, which required the wholesale restructuring of society, even to massive importation of foreign labour. While there were many commercial firms involved in this process, most operated within a structure provided by the state which, in addition to acquiring territory through political means, also invested directly in colonial development through its own companies and through the provision of administrators, garrisons and the like.

As might be imagined, all of this had a far-reaching impact on urbanisation in the new colonies, but there have been relatively few attempts to rationalise what happened in terms of a general theory that encompasses the colonial period as a whole. As noted earlier, by far the most useful is still that by Anthony King who, in his book *Colonial Urban Development*, suggests that colonialism was translated into spatial form through three main intermediary forces: culture, technology and political power. Each of these, King suggests, has markedly dualistic features, being quite different for colonisers and colonised.

Plate 2.1 *Major institutions building in colonies often mirrored contemporary architectural styles in the metropolitan country: A. Mexico City cathedral (source: Katie Willis); B. Saigon City Hall*

A.

B.

Culture

The cultural values of any society, whether social, legal or religious, give rise to a set of institutions which determine much of the physical form of that culture. The various cultural elements of Victorian British society were therefore reflected in the built form of its colonial settlements: churches, theatres, gymkhana and other sports fields (Plate 2.1). Cemeteries too formed part of this cultural dimension. Colonial settlers lived with the threat of death and commemorated their dead in a variety of ways related to their status, religious denomination and artistic taste. These often differed markedly from indigenous ways of dealing with death. Cultural contrasts also extended to the conventional standards acceptable for middle-class colonial housing, such as bathrooms – a factor which is linked to the second intermediary force translating colonialism into physical form; that is, technology.

Technology

The principal contrast between metropolitan and colonised societies in this respect was the more extensive use of inanimate sources of energy, compared to animal and human portage. In terms of transportation this led to the introduction of railways or broad boulevards which devastated the old, medieval street patterns of many existing cities. But the impact was greater than this direct repercussion because European society had already made major socio-spatial adjustments to its own technology – for example, the separation of workplace and residence, and the emergence of functionally specialised buildings. Many pre-existing settlements, in contrast, still comprised multifunctional buildings. New social organisations were also needed to cope with this complex urban system – police forces, transport companies, construction firms and the like. Little wonder that existing urban morphologies were abandoned in favour of new urban districts in which the European technology could be incorporated (Figure 2.2). Such districts were characterised by wide, straight roads and specialised land uses and are almost inevitably focused on the pinnacle of nineteenth-century technological achievement – the railway station – the architectural personification of industrial colonialism (Plate 2.2).

Figure 2.2 *Calcutta: contrasting pre-colonial and colonial urban morphologies*

European suburb of Calcutta

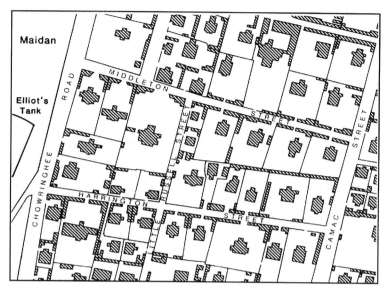

Indian suburb of Calcutta

Plate 2.2 *Malta: the built environment of colonialism*

Political structure

The factor which enabled the wholesale introduction of new technological and cultural values was, of course, the political control exercised by the Europeans. King argues that nineteenth-century colonial society consisted of a two-tier dominant–subordinate relationship of colonisers and colonised. In this context it is important to appreciate that its cities were organised around the control of the economy rather than around the production process itself. There was only limited manufacturing in nineteenth-century colonial cities: large-scale production was usually confined to commercial agriculture or concentrated mining locations. The occupational structure within the cities therefore reflected this political and economic relationship, with a small, metropolitan administrative elite, a small rural managerial group and a large supportive population engaged in the tertiary sector – the production and distribution of consumer goods and services. As the colonial elite also controlled the municipal government, the city could be shaped according to the wishes of a small proportion of its total residents.

Urban development in the industrial colonial period

Colonial influence upon urban systems was, in contrast to the mercantile period, felt at all levels within the urban hierarchy, although variations existed according to time and location. In some areas it is claimed that entirely new urban hierarchies were created – in Malaya for example. In others, such as North Africa, there was wholesale reorientation of urban economic activity from inland trading routes to the new coastal ports. Much of this reorientation was due to the revalued spatial priorities of the new colonialism and was highly selective in its impact. It is this period, therefore, that sees the genesis of contemporary urban primacy as economic and political power is concentrated on certain cities at the expense of others. In settler colonies, such as Zimbabwe, this led to urban development being limited to white-dominated regions of the country – a pattern that has persisted through to the present (Figure 2.3).

Within the colonial cities too, similar patterns of social, economic and spatial segregation were being reinforced. Although manufacturing

Figure 2.3   Zimbabwe: settlement patterns

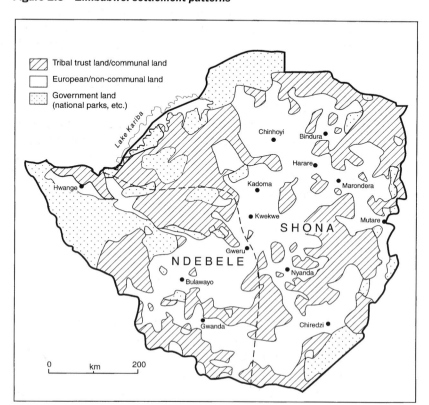

was limited (but not altogether absent) so as not to challenge
European exports, a large commercial and service sector existed to
meet the trading and consumer needs of the colonial power. Much of
this activity was functionally specific to ethnic and class groups within
the city. The foreign component of the trading system was dominated
by Europeans and their institutions, the famous trading firms such as
Jardine Matheson in the Far East. Local assembly and distribution of
the goods involved was often in the hands of expatriate non-
Europeans; only local production, under expatriate supervision, was
the function of indigenous populations. In this system the expatriate
non-Europeans played a pivotal role. Often encouraged to migrate
specifically for this purpose, groups such as the Chinese and Indians in
South-east Asia and East Africa constituted an economic, social and
political buffer between the Europeans and the indigenous masses.
Any haggling over commodity prices in the marketplace thus occurred
between non-European groups and redirected ethnic conflict and
antagonism away from the dominant minority. In some colonial cities
such policies resulted in complex demographic patterns. For example,
by the early decades of the present century only one-third of the
population of Rangoon comprised Burmans; another third was
Indian and the rest divided equally between various Chinese and
European groups. Labour migration also produced considerable
gender imbalance in Rangoon, with the male population being double
that of the female.

This functional distinctiveness was reinforced by ethnic spatial
separation within the cities. Although such separation had been
practised long before the arrival of Europeans, the new colonial
administrators refined and accentuated it by their control of land
ownership. Many of the newly established building and health
regulations were applied only in the European enclaves, which were
effectively separated from non-European zones by the new *cordons
sanitaires* of railway lines, parade grounds, police barracks or race-
courses, reinforced by judiciously placed military cantonments
(Figure 2.4). Thus the urban morphologies of colonial cities were
often dominated in scale and extent by the social, economic and
political needs of what usually remained a demographic minority.

For some cities this overwhelming scale, which was linked more to
imperial rather than local functions, created immense problems after
decolonisation. In Malta, for example, the huge military complexes
that covered the island have proved difficult to incorporate into the
local development process. Some have been converted into residential
or tourist complexes but many others have been reconstructed as new

Figure 2.4 *Proposed ethnic groupings in Singapore City, 1828*

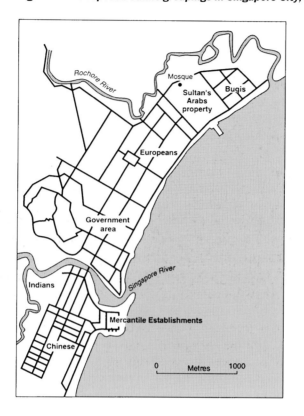

towns, such is their scale in the context of Maltese development (Plate 2.2).

But it would be a mistake to think that colonial cities in the nineteenth century were, as King suggests, characterised dominantly by a sharp dualism between a European and an indigenous way of life, as reflected in architecture, technology and political structures. Very few cities offered such straightforward contrasts, particularly where a buffer expatriate community had been encouraged to develop. Moreover, the meeting of two or more groups frequently produces a hybrid culture and this was certainly true of many colonial cities, being reflected in a wide variety of ways from the fusion of architectural style (Plate 2.3) to inter-ethnic marriage (or at least interbreeding). Furthermore, the European populations in their colonial cities led very different lifestyles from those they had left behind in their metropolitan countries. In the colonies there were usually larger, more spacious houses available, more servants, more leisure and quite different work and social patterns of stratification.

Plate 2.3   *Kuala Lumpur: the hybrid architecture of the railway station*

This contrast was particularly true for the more junior colonial bureaucracy so that the lifestyle of a railway official in Delhi or Bulawayo would be quite different from that of his counterpart in Swindon or Crewe.

Colonial society often exhibited several other kinds of division which further undermine the notion of dualism. The colonialists themselves were fragmented into distinct social groups which resided in quite focused areas of the city – civil lines, military cantonments, railway settlements, commercial districts and the like. When combined with the stratification of indigenous communities, extremely complex social and cultural patterns resulted and were reproduced in the built environment of the city (Case Study D). But perhaps more important than these social and morphological challenges to dualism was the economic structure of colonial urbanisation. This was characterised by interdependence, admittedly unequal but interdependent nevertheless. Not only did indigenous populations 'benefit' from the new employment opportunities generated in and around the colonial city but the colonisers themselves could not have explored and exploited local resources, nor enjoyed such comfortable lifestyles, without the concentrations of indigenous labour brought about by colonial urbanisation.

Finally, it must be remembered, and this point is certainly emphasised by King, that although colonial cities were located at considerable

Case Study D

Delhi: the evolution of an imperial city

The pre-colonial city

When the British arrived in Delhi the city was still the Mogul capital of North India, although it was in the hands of rebels and its population had fallen to about 150,000. Almost all lived within the city wall which was 9 kilometres long and enclosed an area of about 6.5 square kilometres. At the heart of the city were the political, religious and commercial foci of the Royal Palace, the Jama Mosque and Chandni Chowk respectively (Figure D.1). The remainder comprised an organically patterned pre-industrial city of narrow, twisting lanes and mixed land uses – not dissimilar to those of medieval Europe.

Figure D.1 *Mogul residential patterns in Shahjahanabad (Old Delhi)*

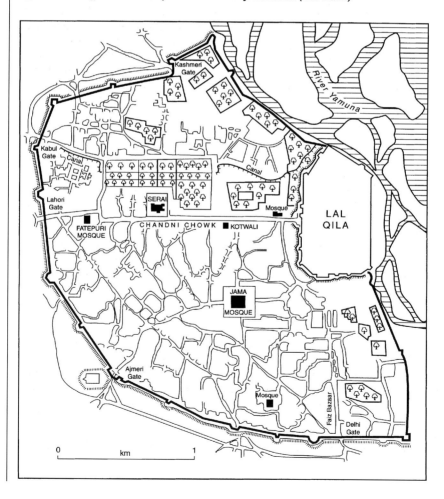

The period of coexistence, 1803–1857

In its early colonial years Delhi was not a major administrative or commercial centre, merely a district military post for the Punjab. There were no more than a few hundred Europeans and, as the situation was very stable, the military cantonment was located to the north-west of the city, with most of the British living in the city itself in an area adjacent to the Royal Palace (where a puppet emperor still reigned) previously occupied by the Mogul aristocracy. Western technology was barely in evidence apart from some architecture, and most of the British lived in the same manner as the indigenous elites. There was little social contact between the races but little conflict either. So, although there was a political dualism, the retention of the court meant that this was not profound and was barely reflected in lifestyle or technology. The principal contrast, therefore, was cultural.

Colonial consolidation, 1857–1911

The pressures of nineteenth-century colonial expansion erupted in India in the mutiny of 1857 in which the Mogul emperor was implicated and dethroned. The consequent sharpening of military control resulted in a more forceful imposition of political power and cultural values. The Royal Palace became Delhi Fort, and a free-fire zone, 500 metres wide, was created around it and the city walls. The military cantonment occupied the northern third of the city and civilians were moved out to the civil lines to the north of the walls where several physically imposing buildings were constructed as institutional symbols of power. The indigenous population was thus confined to the remaining area of the old city which, although crowded, was still functionally and socially sound (Goodfriend, 1982).

Culturally, however, Old Delhi and its population became increasingly isolated as the British withdrew to a distinct area to the north, separated by the military zone, police lines, newly constructed gardens and, above all, by the railways. Poor communications had been blamed for many of the problems of the mutiny, and by 1911 eight separate railways had been linked to the city, further eroding the area available to the indigenous population which had by then reached some 230,000.

Thus the full utilisation of political power had, within the space of some 50 years, transformed the appearance of Delhi, with cultural contrasts spatially accentuated through the medium of a superior military and transportational technology. By 1911 almost a quarter of a million Indians were crammed into 4 square kilometres of the old city, while a few thousand British enjoyed the relatively open spaces of the northern districts.

Imperial Delhi, 1911–1947

It was railway technology that enabled Delhi to be developed as the new centralised capital of a consolidated India – the jewel in George V's imperial crown. It was not until 1921, however, that the physical site was moved southwards to a new location in the crook of the Aravalli Hills. New Delhi was planned by Lutyens and Baker on a vast scale; as early as 1931 it covered 78 square kilometres. This was due to the incorporation of new technological developments, the motor car and the telephone, which in theory enabled such large distances to be bridged. In fact, such technology spread only slowly and erratically and for many years communications depended on people power (i.e. bicycle and foot).

Figure D.2 *New Delhi: the social morphology of imperialism*

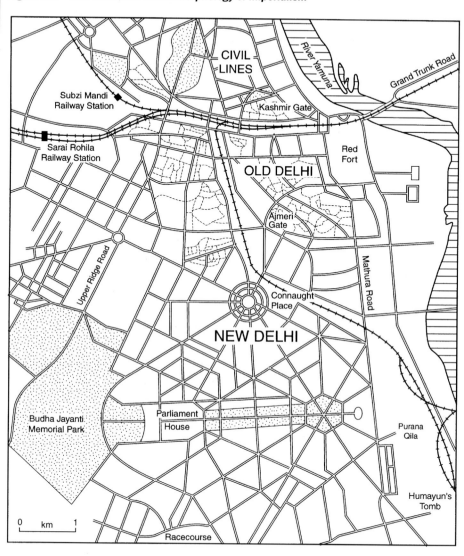

There was no provision for manufacturing growth in the new capital, although the city had acquired several important processing industries such as flour milling. But the most extraordinary example of colonial influence in urban planning was the minute residential stratification of New Delhi into rigid spatial categories based on the infinitely complex combination of Indian, British, military, civil and colonial ranking systems. This resulted in several specifically designated zones (Figure D.2) within which occurred further stratification of size of house, garden and facing direction in relation to the social Mecca of Government House.

The old city received some improvements to drainage and water supply systems but, as in most colonial cities, urban planning was primarily reserved for the expatriate zones. Old

Delhi became even more physically and socially isolated by more gardens and a spacious new business district. As a result in-migration led to massive overcrowding, deterioration of the urban fabric and overspill into rapidly growing squatter areas to the west and east of the old city. Not surprisingly, by the Second World War death rates in Old Delhi were five times those in the new capital.

Independence did not change this contrast very much. Ten years after this event Old Delhi still contained 60 per cent of the city's population at an average density of 41,300 per square kilometre. But even this was but a foretaste of what was to come later as the capital's population quadrupled to its present size of almost 8 million. Old Delhi still suffers constant water and electricity shortages, and human waste continues to be removed by basket. It is, however, unfair to blame the present situation on the inadequacies of Delhi's contemporary planners. The roots of the present situation lie firmly in the segregationalism of its colonial past.

distances from Europe, they played a crucial role in the global economy. Although not producers in their own right, colonial cities were controllers of production and redistribution and as such were crucial points within the circuit of reinvestment and money capital that characterised the global economy of the nineteenth and early twentieth centuries. Without them the economy of industrial Europe would have collapsed. It is, perhaps, this pivotal economic role of colonial cities that underpins the political, technological and cultural processes which King emphasises as major shaping forces. King's model is, therefore, a useful pointer to the processes that moulded colonial urbanisation, but underplays, perhaps deliberately so, the economic links between and within colonial cities and overemphasises their dualism.

Late colonialism

In many ways most contemporary images of colonialism are those of the inter-war or immediate post-war years. The writings of Somerset Maugham or Karen Blixen, and related films, conjure up images of sweating white-suited men and elegant, broad-hatted memsahibs being served pink gins or iced tea by smiling, courteous 'boys'. The reality for all but a fortunate few was, of course, quite different. Two horrendous global wars and an intervening economic recession resulted in erratic demand for the primary products of the colonies, and profits became unreliable. In most areas this resulted in the introduction of efficiencies through economies of scale and smaller producers were ousted by land reforms and mechanisation – the development of capitalism in depth.

This began to accelerate the drift of rural labour to the towns where, to some extent, it was incorporated into an expansion of domestic service and factory-based production, but as yet this was limited in scale. The growth in indigenous populations appears to have had relatively little impact on urban planning. Although squatter settlements began to appear, the principal morphological change occurred in European quarters where colonial architecture and land-use planning reached their most grandiose form. It was in the late colonial period that the huge institutional legacies of city halls, universities, banks and the like were bequeathed to the contemporary world (Plate 2.4).

The pressures of population growth in many of the larger cities eventually forced the authorities into the realisation that a more systematic form of town planning was needed. Up to this point Europeans had been ambivalent about urban population growth. On the one hand, they needed more labour; on the other, they did not want 'excessive' numbers of indigenous people posing what was seen as a threat to their physical and social health. Decisions on urban administration had by now largely passed to the colonies themselves, partly as a result of increased European settlement; the consequence was a distorted planning process which was dominated by Europeans and their interests. Thus the 'garden city' movement which had strongly influenced British town planning in the early twentieth

Plate 2.4 *Johannesburg: the University of the Witwatersrand, an architectural legacy of the late colonial period*

century was used in the colonies primarily to accentuate the segregation that had been consolidated in the previous century. The *cordons sanitaires* around the European districts began to diversify from military barracks or railway land into parks, golf courses and the like. In some colonies such planning was eagerly incorporated into an emerging philosophy of racial apartheid (Christopher, 1988). More and more, therefore, the cities were becoming increasingly detached from indigenous life, whether rural or urban. One Malaysian writer, Lim Heng Kow (1978), described Kuala Lumpur in the 1930s as 'a city of European government buildings, European and Oriental banks and businesses, and Chinese and Indian traders or workers, with occasional residual patches of the otherwise submerged Malay world'.

Although the late colonial period saw relatively little structural change in the city, there was in certain colonies an important demographic transformation. This was the accelerated migration from the European recession of large numbers of blue- and white-collar workers. British migrants could settle in largely British communities in the dominions of Canada, Australia and New Zealand, but the French, Dutch and Belgians still comprised minority groups, albeit quite extensive, in their colonies in South America, Africa and Pacific Asia. Increasingly such migrants began to infiltrate lower white-collar or retail occupations not hitherto favoured by Europeans. Even in British Rangoon well over one-quarter of the European population in 1931 could be classified as traders, shop assistants or unskilled/semi-skilled labour. One unforeseen, but eventually momentous, consequence of this expatriate influx was that the slowly expanding group of educated indigenous residents found it difficult to break into middle-income occupations in either commerce or administration. Their subsequent drift to a Europe driven by political extremism quickly transformed many into political activists who eventually played important roles in the independence struggles in their own countries. It also led to difficult, brutal and often bloody confrontations over decolonisation in the French and Dutch colonies in modern-day Algeria, Vietnam and Indonesia where European settlement had been extensive.

Early independence

The 1950s and 1960s saw independence spread rapidly throughout most of Asia and Africa, although in those areas where the inter-war European settlement had been substantial, such as Algeria or Java, decolonisation was often slow and bitterly contested by colonists who had little to gain from returning to a war-ravaged Europe. Once the colonial powers had departed there was a great surge of indigenous peoples into their cities, attracted by the prospect of jobs in the lucrative administrative and commercial positions from which they had been excluded. However, in the early years of independence such jobs were relatively few, given the persistence of European control within commercial enterprises and the sluggishness of demand for primary materials from a shattered European industrial economy.

Ironically one of the major problems in Europe was shortage of labour – of which the cities in the newly independent Third World had far too much. The result was an encouragement to this 'surplus' labour to move to employment in north-west Europe. Initially the movement was from former colonies to former metropolitan powers (for example, Algerians to France, or Indians to Britain), but it soon spread to many other poor countries, particularly those around the Mediterranean Sea. For European industrialists and governments this migrant labour was an economic lifeline. It was abundant, non-unionised, cheap and docile (being easily threatenable with deportation), and for two decades the economic miracles of several West European countries were firmly built on the back of this exploited labour. For their part, Third World governments were only too glad to encourage such a move since it helped decelerate urban population growth, increased foreign exchange revenues through savings sent home and, they hoped, trained some of their workforce.

The build-up of workers in Europe was rapid and highly concentrated. By the late 1960s West Germany and France alone had some six million guest-workers between them, with particular concentrations in industrial cities and in the 'dirty' or 'monotonous' types of employment. But for the donor countries there was comparatively little reward. They lost large numbers of their younger and most trainable workers; few migrants received useful skills or training; and the remittance money was mostly used to finance small consumer businesses in the big cities on the return of the migrant – thus accentuating the original problem (see Findlay, 1994).

By the 1970s the migrants were becoming less welcome. A growing recession left unemployed Europeans resentful of migrant workers; the migrants themselves were becoming more unionised, less docile and demanding higher wages and better living conditions. By the middle of the decade most European countries had tightened their labour immigration laws, although dependants continued to migrate and by the mid-1990s some 8.5 per cent of the German workforce was of Turkish origin. This build-up of minority groups has undoubtedly fuelled right-wing politics during subsequent years, with many unfortunate consequences in several European countries where even long-term migrants still lack political rights.

Meanwhile, in Third World cities themselves there was little immediate economic change following independence. The new nations continued to be dominated by the same commercial and trading activities in primary exports and manufactured imports and by the same expatriate firms. It was simply another form of colonialism. Socially, the major transformation was the emergence of a huge subsector of urban poor who were unable to acquire waged employment and who were excluded by their poverty from acquiring adequate housing, education and health care for themselves and their families. Such households began to look inwards for their survival and create their own economic system in which meagre incomes were earned in a variety of illegal and semi-legal activities and then spent within a production system that specialised in the creation of a consumer market apparently operating outside the formal supply network. This has come to be known as the 'informal sector' or 'petty-commodity production' and is examined in more detail later in this volume.

In the late 1960s and early 1970s the world economic system began to change drastically with regard to its incorporation of Third World labour. In Europe productivity had declined as wages rose and environmental concerns too began to increase production costs. Accordingly, many European and North American companies began to shift their points of production into the cities of the Third World where cheap labour still existed and could be guaranteed by authoritarian governments that were reliant on the West for political support.

Over the last few decades, these investment funds have increasingly gone not directly into Third World factories but into multinational corporations, such as Cadbury-Schweppes or Lonrho, who are operating or willing to operate in developing countries. Why should this have occurred? In the first place, as we have seen, the costs of production had risen in Europe – not only for wages but for rents and raw material imports too. Second, labour is cheap in Third World cities as a result of accelerating rural–urban migration (itself the consequence of capitalist penetration of the countryside). Third, the presence of the large informal sector, a reserve army of labour, helps keep down the demand for wage increases. Fourth, the recent advances in technology have enabled fragmentation of the production process from management. The advent of email, telex, satellite links and containerisation has meant that it is possible for the labour-intensive parts of the production process to be located in Third World cities, while retaining specialist management, research and development in the metropolitan country. Finally, this entire process has been encouraged by international agencies and national governments all anxious to bring employment to the burgeoning cities of the Third World in order to forestall possible political instability.

The impact of these changes on the cities of developing countries has been varied and complex (and will be examined in more detail in subsequent chapters). Perhaps the most important point to notice here is that rapid economic growth has been extremely selective, so that only some six or seven countries, such as South Korea or Taiwan, can be said to have rapidly expanded their industrial economy in the early years of this investment process. However, as labour costs rose in these locations, so the multinational corporations, together with local producers, began to look further afield for new supplies of cheap urban labour, and new industrial producers have emerged, such as Malaysia and Thailand. In general it is the already large cities, with burgeoning populations and infrastructural facilities, that have received the brunt of the investment. This in its turn has induced a third wave of migration into such cities – not all of it being from within the country concerned.

The social effect has been considerable in terms of class formation. A waged proletariat of varying importance has appeared in many cities. Being more privileged than many of their fellow citizens they can form a relatively conservative labour group, although in some cities agitation for better working conditions has occurred. In contrast, the informal sector has continued to proliferate and many outside observers fear that continued deprivation and frustration, particularly in the wake of structural adjustment and economic downturns, may induce a demand for more radical change. Moreover, in some countries where varied ethnic groups have been attracted into the city, in both the colonial and post-colonial periods, another dimension of urban instability has appeared (Dwyer and Drakakis-Smith, 1996). A final area of social change has been the unprecedented incorporation of women into the urban workforce, a phenomenon very different from the early years of independence.

All of these changes have been undertaken in most cities with the assistance and co-operation of national governments. The role of the state in taking over the management of capital cities from local authorities has been a crucial step in facilitating the penetration of foreign investment. In their eagerness to encourage growth many governments spent heavily on creating a built environment of modernisation in their cities, borrowing money to construct new airports, conference centres, free trade zones and the like. Many, of course, over-borrowed and found themselves in debt to international banks and agencies. Consequently many governments have had to adopt a series of economic adjustment programmes imposed upon them by the World Bank or the International Monetary Fund as a precondition for further loans. An important component of these structural adjustments has been the reduction of government expenditure in areas such as social welfare. These cutbacks have hit urban populations particularly hard, worsening the shortfall that already existed in the provision of basic needs.

Not surprisingly, it is this range of recent economic and social changes and their consequences that form the focus for the remainder of this book on contemporary Third World cities. What this historical chapter has done is to emphasise the fact that we cannot hope to understand their true nature without appreciating the global setting of their evolution.

Key ideas

1. Colonialism was not a uniform process and varied enormously through space and time, and in relation to the cultures involved.
2. The urban impact of mercantile colonialism was very limited.
3. The impact of industrial and late colonialism was extensive both on urban hierarchies and on individual city morphologies.
4. Culture, technology and political power transformed economic objectives into urban form.
5. The major changes to this situation did not come with independence but following transformation of the world economic system in the last quarter of the twentieth century.

Discussion questions

* Describe the main features of mercantile colonialism and explain how they affected colonial settlement.
* What were the major forces transforming economic objectives into urban form in the nineteenth century? Illustrate your answer with reference to a case study.

* What were the major contrasts between urban growth in the late colonial and early independence periods?
* In what ways did independence affect urbanisation in the developing world?

References and further reading

Christopher, A.J. (1988) *The British Empire at its Zenith*, Croom Helm, London.

Dixon, C. and Heffernan, M. (1991) *Colonialism and Development in the Contemporary World*, Mansell, London.

Dwyer, D. and Drakakis-Smith, D. (eds) (1996) *Ethnicity and Development*, Wiley, London.

Findlay, A.M. (1994) *The Arab World*, Routledge, London.

Goodfriend, D.E. (1982) 'Shahjahanabad: tradition and planned urban change', *Ekistics*, 49: 297–298.

King, A. (1976) *Colonial Urban Development*, Routledge & Kegan Paul, London.

King, A. (1990) *Urbanization, Colonialism and the World Economy*, Routledge, London.

Lim Heng Kow (1978) *The Evolution of the Urban System in Malaysia*, PUM, Kuala Lumpur.

Lowder, S. (1986) *Inside Third World Cities*, Croom Helm, London.

McGee, T.G. (1967) *Southeast Asian City*, Bell, London.

Nath, V. (1993) 'Planning for Delhi', *Geo Journal*, 29(2): 171–180.

Taylor, P. (1985) *Political Geography*, Longman, London.

③ Urban population growth

- Introduction
- Migration to the city
- Explaining urban migration
- Migrants and migration
- The impact of migration
- Government responses to migration
- Natural population growth and the city

Introduction

The previous chapters have established that Third World urbanisation has undergone tremendous acceleration and change over the last two decades. This chapter will examine the two main components of this population expansion – migration and natural growth – not only to re-emphasise their relative importance (this is discussed in the broader demographic setting by Parnwell, 1993, and Findlay, 1994, in this series), but also to try to tease out of the discussion some of the forces which have influenced individual families in their decisions on these matters. In this way we can place some of the general economic and political issues raised in the previous chapter into the household context. It is important, therefore, to look beyond the demographic dimensions of urban population growth and to consider the ways in which the changing qualitative composition, such as age, ethnic and skill structures, together with access to basic needs, affect urban sustainability (Gould, 1998).

The two components of urban population growth vary in relative importance through space and time, but in general, migration is more important in the early stages of urban population growth when the proportion of national population living in towns and cities is low (Figure 3.1). As the urban proportion rises, so does the contributory role of natural growth, although only up to a certain point. Beyond this point, which is related more to the demographic cycle than to the absolute size of urban population, urban fertility begins to decline

Figure 3.1 *The relative importance of migration and natural increase in urban population growth*

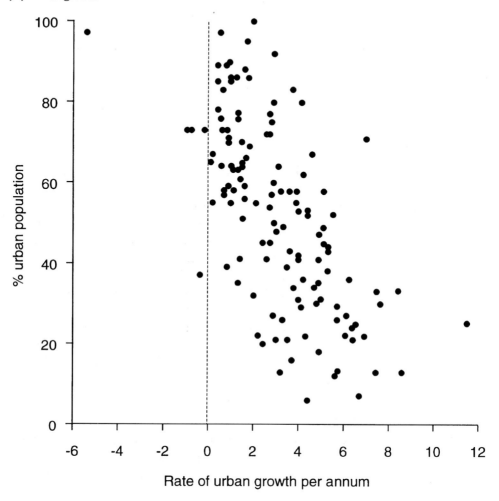

High rates of urban growth are usually the consequence of a high proportion of urban migration in total growth and often relate to middle ranking or low levels of urbanisation overall

and migrational growth once again becomes more important, albeit at a drastically reduced level. So in developed countries population movements rather than natural growth dictate overall population change in most cities.

In recent years, the migrational component of urban populations has become much more difficult to identify with the rise of circular migration, a temporary move to the city by those who retain a formal, census residence outside it. Such long-term commuting has increased markedly in recent years as transportation has become cheaper and more regular. This has had important implications for rural–urban linkages which have become much more fluid. Production and consumption by individuals need not occur in the same place, thickening and deepening the nature of rural–urban flows.

Migration to the city

There is a considerable literature on migration to towns and cities, most of which seems to concur with Todaro (1994) that migration is primarily motivated by perceived economic opportunities in the city. A strong association between economic and urban population growth has emerged from such theories. However, the precise nature of this link varies over space and time, and there are many contradictions. For example, in a large number of countries in Africa rural poverty has induced urban migration despite the lack of economic growth in their settlements. In contrast, in other poor countries, such as Bangladesh, this massive urban shift has not occurred. Similarly, although rural to urban movement of capital and labour has been an important component of economic growth in South Korea and Taiwan, this has not happened in Thailand where unprecedented economic growth is occurring in a country which is still overwhelmingly rural.

It must be noted at the outset that increased economic pressure on the rural household does not always result in urban migration within the country concerned. As Mike Parnwell (1993) has observed, 'not everyone needs to migrate, not everyone wants to migrate and . . . not everyone is able to migrate'; some governments, though, have successfully deflected migrants to other rural areas or to more developed countries. However, despite these observations, by far the most common response to rural poverty is internal migration in towns and cities.

Until recently it has been assumed that the largest cities have borne the brunt of this movement – an assumption apparently verified by the sheer numbers crowding into capital cities. However, it is instructive to remember that two-thirds of the urban population of the South live in cities of under one million. Moreover, there are signs that in many large cities migration is of reducing importance. In Latin America, certainly, there has been considerable evidence for some time that migrational growth comprises only one quarter to one-third of total population growth for most large cities. For example, in Mexico during the 1980s the proportion living in urban centres of between 15,000 to one million rose from one-quarter to one-third of the total. Even in Argentina 47 per cent of the urban population lives in settlements of less than a million. Whether this is the result of rapid natural growth, disaffection with the opportunities available in cities with already huge populations living in poverty, or the increased attractions of intermediate cities, is difficult to judge given the relative lack of information. Whatever, the plain fact is that a large proportion of urban residents in the Third World live in relatively small settlements; at least 40 per cent live in urban centres of 100,000 or less. As more reliable data on urban hierarchies as a whole are made available, it is becoming clear that smaller towns and cities have also been drastically affected by migration, perhaps even more so as they are centres of both in- and out-migration (Plate 3.1).

Such upheavals are not always reflected in the collated data available on this topic because the double flow of migration tends to balance out, giving the impression of a stagnant population (Table 3.1). One must also be cautious about these data because some capital cities are still relatively small and will appear in the intermediate categories, although they are the final migrant destination within their country. In addition, as cities grow, they will change categories so that the most

Plate 3.1 *Kenya: rapid population changes – small towns, too, can experience severe problems in coping with migrational growth*

Table 3.1 *Average annual rates of growth and city size in developing countries*

	1950–60	1960–79	1970–80	1980–90
20,000–99,999	4.9	1.2	0.1	–0.5
100,000–499,999	4.2	3.5	4.4	2.6
500,000–999,999	3.5	4.1	5.4	4.4
1 million or more	8.2	5.4	5.6	5.6

Source: D. Rondinelli (1982) 'Intermediate cities in developing countries', *Third World Planning Review*, 4(4).

rapidly expanding centres will sooner or later work their way into the top rank, taking their growth rates with them. But perhaps the most important criticism is that such tables do not incorporate the 'hidden' urbanisation of circular migration in which both short- and long-term commuters retain their rural base and are not revealed as migrants in census data. Their presence in the city does, however, increase even further the demand for employment and facilities of all kinds. (See Case Study E.)

Explaining urban migration

Mike Parnwell (1993) has divided the reasons why people move into three groups – macro, meso and micro; this section examines the first two of these. Macro reasons encompass the fact that most migration is related to uneven economic development in that it results in shifts of people from poor to better-off regions. In most instances such uneven geographical development has been the result, deliberate or otherwise, of national or international development policies – from colonial through to contemporary times. However, as Parnwell cautions, the simple existence of such imbalances does not, in itself, provide sufficient reason for migration. Migrants are individuals and make their decisions on highly personalised grounds and not in response to national trends.

Within the meso-level context of the rural areas themselves there are several factors which can stimulate migration. One of these is investment into large-scale and/or commercial agriculture which is often intended for export. The result is increasing landlessness and relatively few farm and non-farm jobs. In China, for example, the consequence of recent economic change, together with the 1966–1976 baby boom, has been to increase the number of surplus rural workers to 200 million. Poor service infrastructure and basic needs provision

Case Study E

Intermediate towns and cities

It is estimated that some two-thirds of the urban population of the developing world live in cities of less than one million people. Of course, intermediate urban centres cannot be defined by population alone and in many countries a settlement of one million would constitute a very large city indeed. More important as a definition are the relative roles that intermediate cities play in sectoral and spatial development, and many attempts to promote the growth of small towns have foundered on the misplaced assumption that regional and functional roles can be separated. The demographic role of intermediate settlements usually varies according to local circumstances but can often be quite complex. For example, Miriam Grant's work in Gweru (Grant, 1995), the fourth largest urban centre in Zimbabwe with a population of just over 100,000, reveals that only about one-third of its migrants come from villages and rural areas, the remainder are from other urban centres, most of which are larger than Gweru itself, notably the two major cities of the country, Bulawayo and Harare.

In the past, intermediate settlements were largely regarded as conduits through which rural resources, both material and human, flowed up the urban hierarchy to the core of the political economy. More recently, small towns have been viewed more sympathetically as components of a programme of decentralisation of infrastructure and services, particularly by international agencies and donors (Baker and Pedersen, 1992). What is still missing, however, is an approach which seeks to base development policies on the needs of those who live in and around these settlements – a bottom–up approach to urban growth. At present such approaches fly in the face of most development strategies, which seem to emphasise the inevitability of high levels of urban concentration within the regions of more rapid economic growth. The consequence has been a neglect of non-farm development opportunities in small and intermediate towns which has led to the steady build-up of 'surplus' rural populations to sustain or to accelerate rural–urban migration to large cities. As noted elsewhere, recent estimates suggest there will be some 200 million 'surplus' rural workers in China by the next decade. The consequences of a shift of this magnitude into the already congested mega-cities of China will be enormous. People who once grew food will now need to purchase it in the city and so become part of the urban cash economy, placing further burdens on rural producers and urban administrators. Such pressure exists throughout the developing world and suggests an important role for intermediate centres.

in rural areas also contributes to urban migration, as does the steady, continuous and cumulative environmental disruption and deterioration. Parallel developments in transport have facilitated this movement (Simon, 1996). However, the link between rural fertility, population pressure on resources and migration is extremely complex and depends very much on the local situation (see Lockwood, 1995, for further discussion).

Much of the migration from rural to urban areas is undoubtedly

Table 3.2 *Kenya: rural and urban contrasts, 1993*

	Urban residents (%)	Rural residents (%)
No education (aged 6 or above)		
Female	13.5	29.1
Male	7.0	18.2
Household possessions		
Radio	67.7	48.1
TV	22.0	2.4
Electricity	42.5	3.4
Piped drinking water	55.8	10.7
Flush toilet	44.9	1.6
Health of children		
Mortality rate (under 5)	75.4	95.6
Infant mortality rate	45.5	64.9
Underweight	12.8	23.5
Maternal health		
Receiving tetanus jab	92.9	88.8
Receiving formal prenatal care	97.6	94.5
Receiving formal delivery care	77.6	38.2
Total fertility rate	3.4	5.8

generated by the perception of improved opportunities and facilities (Table 3.2) in the city, rather than simply being the consequence of increased rural population growth. Indeed, in some countries migration to the city has left important rural labour shortages, particularly at crucial times of the farming year. To be sure, where rural population growth generates increased pressure on limited rural resources, such as cultivable land or on waged work, the impetus to migrate may be difficult to resist, particularly when opportunities in the city are perceived to be so much better (but are seldom achieved).

Clearly migrants are realistic about what they can expect, in terms of life in the city, and yet they come in ever-increasing numbers. The reason is that for many life does improve in the city, or at least prospects for a better life improve. In this context time-scales are important and many rural migrants are prepared for a long struggle before they or their children can reap the benefits of their hard work.

Of course not all will experience such improvements and here we must make another distinction between those migrants who are drawn to the city by its better prospects and can take advantages of these, and those who were forced reluctantly out of their home areas by a variety of push factors and who are ill-prepared for life in the city.

The apparently clear-cut nature of such contrasts has willed many economists into producing models aimed largely at predicting migration rates through rural–urban income differences (Todaro, 1994). At their simplest these models reduce migration to a straightforward choice between a traditional, backward, rural way of life and a modern, industrialised lifestyle in the city. But this contrast rarely exists in reality in the contemporary Third World. As a result of improved transportation and other communicational developments, particularly the spread of radio and television, urban values and goods have long since penetrated even the remotest regions; a process which is reinforced by return migration. In parallel, traditional rural values have moved into the city with the migrants and are retained through the tendency for people of similar backgrounds to cluster together. Indeed, in many ways it is the blurring of distinctions which has served to encourage migration because there is not now such a drastic change of lifestyles involved. In Vietnam, for example, although the average urban income is more than double that of rural areas, one-third of urban dwellers still receive less than the average rural income (Drakakis-Smith and Dixon, 1997). Such rural–urban social complexities are, in some cities, accompanied by economic integration as urban-orientated but space consuming industries, such as brick-making, are forced to move out by rising land prices and to relocate in nearby rural settlements, binding them tightly to urban consumer and labour markets, and helping create the complex mega-city regions found in many parts of Pacific Asia.

Migrants and migration

At the micro level it is clear that migration is not an individual affair, even when only one person migrates. Usually it is the consequence of a collective decision involving immediate and extended family members, together with any other friends who may be able to provide relevant comment. It is seldom a hasty decision but rather one made on accumulated information on the perceived opportunities available in the city. Moreover, as the migrant may not find immediate work in the city, there may be relatively substantial 'support costs' required during the transition period. So-called 'spontaneous' settlement is, therefore, a clear misnomer.

Although the decision to move is likely to be a family affair, migrants themselves tend to be self-selective because of their personal attributes. However, the most 'suitable' member of a household may still be uneducated and unprepared for the problems of life in the city and fail to obtain an acceptable income, even within the meagre expectations they have. As time progresses those migrants who have remained in the city are joined by others less able and more dependent, increasing substantially the costs of feeding and housing the migrant household. It is often at this point that migrant households shift from small, rented shelter near the city centre to expanding squatter communities where there is more housing space at lower cost (see Chapter 6).

The physical move itself is heavily dependent on personal contacts, occurring within what anthropologists call a 'spatially extended social field' of kin, tribe or community links which is held together by traditional values related to mutual obligations. Once in the city, it is the same family or friends that usually provide early accommodation and assist in obtaining work. For many years a common pattern was assumed to characterise the bulk of migration throughout the Third World; that is, most migrants were adult males, the vanguard of a later family shift, who sought work in a series of step-like moves up the urban hierarchy at increasing distance from their point of origin. In other words, they would first migrate to a nearby small town, learn some skills and establish an income, then move on to a larger settlement with a greater range of opportunities. Later moves might be by sons or grandsons, with families catching up from time to time. Such stereotyped movement no longer dominates migration patterns. Much of this is due to new transport developments which have made longer-distance travel easier and cheaper, often at the expense of local destinations. For example, Roi Et, a small town in north-east Thailand, has several buses a week making the 900-kilometre trip to the capital Bangkok – a much better service than to the regional capital of Khon Khaen. As a result, in many parts of the developing world migrants tend to make a direct, long-distance move to a large city, whose long-distance bus termini are usually hives of activity surrounded by services of all kinds (Plate 3.2).

In widespread areas of the Third World transport improvements have also accelerated circular migration, which is really long-term commuting, with the migrant retaining a rural home but moving to

Plate 3.2 *Harare: Mbare long-distance bus terminus*

the city for weeks or months at a stretch. This is, of course, a common-sense response by the poor who are attempting to obtain the 'best' of both world's by reducing expensive living costs in the city and retaining rural land revenues or food sources: a process of 'earning in the city, spending in the village'. Many urban employers welcome this trend too, because it keeps wages low since circular migrants often live in very cheap lodgings. The main problems are faced by the urban authorities whose city populations are swollen for most of the year by such migrants and who must meet their housing, food, transport and health needs when these arise. In addition, considerable social and economic transformation often occurs in the zones immediately adjacent to such cities, with rapid growth of informal activities of all kinds – most of which are beyond the control of the urban authorities and produce a chaotic and environmentally degenerative environment, with squatter settlement or illegal industry often cheek-by-jowl with golf courses or airports.

Closely linked to the shift to circular migration has been another marked change in trends which is the rapid growth of female migration to the city. Some countries have always had comparable levels of female migration, particularly those where rural land holdings are in tenure farms and where there is little economic incentive for the family not to accompany the male migrant. However, over the last decade there has been a massive upsurge in the demand for female labour in some cities following the expansion of manufacturing production in many parts of the Third World. However, the incorporation of women into the migrant stream in response to the availability of increased opportunities in factory work is still limited to a small number of rapidly industrialising countries (Figure 3.2). The impact of female migration to the city can be quite far reaching, not only in the city itself but also in the rural areas where economic and social life has often been dramatically affected by the loss of the hardest working member of the community.

Figure 3.2 *Women as a proportion of migration streams*

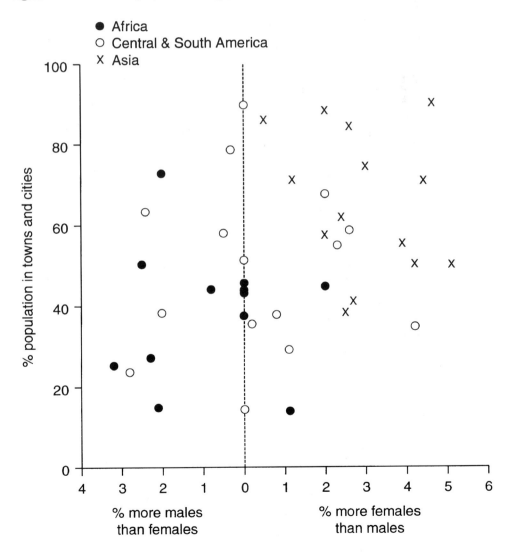

The impact of migration

The loss of young labour is usually assumed to have a deleterious effect on rural life, particularly in economic terms, so that eventually more migrants move away because of a widening gap in the quality of life between rural and urban areas. There is, however, an alternative position which argues that the loss of surplus labour has only a limited impact. The various reasons suggested for this are first, that continued population growth in the context of fixed land availability soon compensates for out-migration; second, that much rural labour is underemployed, so that out-migration leads to a more efficient use of remaining labour; and finally, that much out-migration is circular and occurs during slack periods in the agricultural calendar. Not all of these arguments apply in all cases and, indeed, articulation between the agricultural seasons and circular migration soon declines as migrants obtain employment with more formal working requirements.

Nor can we say with any certainty that migrational impacts relate more to males than females. As we have seen both are in demand in the cities and the introduction of agricultural mechanisation, whether as a cause or consequence of migration, affects the genders differently in different countries – although new techniques in post-harvest activities do tend to affect women more. However, we must not confine discussion to economic matters. Rigg (1996), for example, suggests that in South-east Asia much of the impact of rural to urban migration is cultural and social. Slack periods in the farming calendar were/are times for festivals, fairs and courting. The migration and absence of young people from such occasions devalues them, while in turn the new cultural values of the migrants makes them look down on these 'traditional' activities and on those who remain in the village. This is particularly the case with female migrants who may return wealthier and more worldly-wise than their prospective partners. The migration process also gives rise to households with complex livelihood strategies based on both rural and urban activities, particularly those in peri-urban areas. Often, in such circumstances the land usage changes, perhaps to more valuable urban-oriented crops, even to keeping land idle as a speculative investment holding for sale at some appropriate point in time.

The impact of migrants on the city is, of course, quite a different matter and ironically is often seen in adverse terms as leading to an over-supply of labour that sediments to the bottom of the labour market (Table 3:3), while at the same time imposing enormous demands on the city in terms of shelter, education, health care and the

Table 3.3 *Characteristic roles of rural migrants in the urban economy in Asia*

....

Informal sector

- Vendors, petty traders: 70 per cent of street traders in Manila are migrants to the city, and about half are temporary sojourners

- Garbage pickers (scavengers) and junk buyers in Hanoi: more than 50 per cent come from Xuan Thuy district in Nam Ha province, south of the capital

- Drivers of becaks (Indonesia), saam lors (Thailand), cyclos (Vietnam, Cambodia): all variations of tricycle taxis. Up to 40 per cent of the transportation in Bangkok is manned by seasonal migrants

- Foodstaff operators

Casual sector

- Construction workers in Bandung and Jakarta: the 'majority' of tukang (manual workers) in the construction industry in Jakarta are not natives of the capital; most come from rural areas of Central and East Java and were formerly wage labourers in wet rice agriculture

- Security guards

- Domestic servants

Formal sector

- Workers in labour-intensive, low-skilled, export-oriented manufacturing (garments, footwear, etc.): 55 per cent of workers in a large export-oriented garment factory on the outskirts of Bandung, Java commute daily from rural areas

....

Source: Rigg (1996).

like. These issues are more properly dealt with in later sections of this volume but have stimulated debate on the attitudes and responses of the state to the migration process itself. A brief overview of these responses is, therefore, worthy of discussion.

Government responses to migration

Full discussion of the responses to migration lie beyond the scope of this particular text and can be found in other books in this series, particularly those by Mike Parnwell (1993) and Chris Dixon (1990). However, it is worth reviewing the range of options open to Third World governments and drawing some general, if brief, conclusions about their popularity and success, because these have had considerable impact on attitudes towards the second component of urban population growth: natural increase. It must not be assumed that all governments share a common enthusiasm for curbing migrational growth – after all, urban industrial expansion depends on cheap labour which in turn is linked to continued population growth. Moreover, easing population pressures in the rural areas may help to balance uneven regional development. However, international agencies are often more concerned with the threat to political stability posed by the continued build-up of urban poor. As Robert McNamara, former President of the World Bank, once warned, 'if cities do not begin to deal more constructively with poverty, poverty may begin to deal more destructively with cities'.

Of course, the most appropriate responses to rural out-migration must be those that directly address poverty and underdevelopment in the countryside itself. Programmes of this nature involve such matters as land reform, infrastructural investment, generation of non-farm employment and rural inputs into the planning programme. Unfortunately these tend either to be very expensive or to be associated with socialist co-operative movements, neither of which appeal to poverty-stricken capitalist governments. Many have, therefore, resorted to other less economic responses designed simply to prevent population movement into the already overcrowded cities. Some of the measures adopted in an attempt to curb urban migration have been variations on the old South African pass system, designed to prohibit urban residence to those without official permission. Without draconian police enforcement, such systems have been abject failures, spawning corruption or simply being ignored by migrants, many of whose informal-sector activities are already illegal anyway.

Slightly more 'successful' have been attempts to deflect or redirect migrants away from large cities to other reception areas. Where newly developed or developable land exists, as in Indonesia (see Case Study F) or Malaysia, some success may be achieved, although in small numbers in relation to total migration flows. More effective, at least for a short while, has been the encouragement of potential migrants to move directly overseas to labour-short urban centres abroad. As discussed earlier, this was a notable feature of certain Mediterranean and former European colonies in the 1960s and 1970s, and over the past two decades has increasingly come to characterise the developing world per se as people move to the more successful cities in search of better economic opportunities (see Case Study G).

Case Study F

Transmigration in Indonesia

Between 1975 and 2000 Indonesia's population grew from 135 million to an estimated 212 million. Most of this increase occurred on the island of Java where 60 per cent of the country's population are crammed into 7 per cent of the land area. Although it is a fertile island, Java cannot sustain annual increases in the region of two million people, and almost half of its population live on or below the poverty line. This overpopulation has long encouraged both the colonial and Indonesian governments to alleviate the situation through programmes of planned population shifts. These began in 1905 when the Dutch began to encourage labour movement to their new plantations in Sumatra, but in the four decades of its operation this scheme transferred only 200,000.

Since independence in 1950, over five million migrants have participated in the transmigration programme. Most moved during the 1980s and the programme has faded substantially in the 1990s, largely because of its shortcomings. The impact of transmigration has been minimal, for example, on the sending areas of west and central Java where natural increase has quickly replaced those who have moved, and these provinces remain amongst the most densely settled parts of Asia, reaching over 500 persons per square kilometre over large areas. Poverty too is still double the national average.

In the receiving areas the situation is very disappointing. Many of the migrants were the less enterprising and skilled from their home area and were settled on tiny farms, usually 1 or 2 hectares, in environmentally marginal areas. Inappropriate methods and crop types led to land deterioration and official support was slow and desultory (Sage, 1996). These environmental problems were a major factor in the World Bank's decision to cut its financial support for transmigration in the 1990s.

Figure F.1   *Indonesia: directions of transmigration*

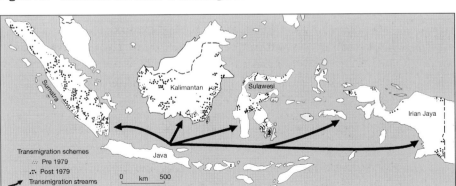

However, there was another factor undermining the programme. This was the substitution of political motivation for the original demographic/economic rationale. Essentially, the

Indonesian government began to use transmigration of Javanese people as a strategic means of securing ethnically unstable border areas (Figure F.1). This is particularly noticeable in Irian Jaya, the western half of the island of New Guinea, where the indigenous population is Melanesian not Malay. West Irian contains 20 per cent of Indonesia's land area but only 1 per cent of its population and there are plans to settle one million Javanese into the indigenous population of 1.6 million. In fact Javanese settlement has reached only about 20 per cent of this target but it has been concentrated into towns where it comprises two-thirds of total population and where Javanese dominate administration, business and commerce. This economic and cultural (Islam and Christianity) antagonism has been having the reverse effect on the region to that which the Indonesian government wanted. Transmigration in Irian Jaya has led to militant demands for separatism rather than promoting national unity. These have increased with the weakening of central government controls in the late 1990s.

Case Study G

International labour migration

Before the 1970s most international movement of labour was from developing to developed countries, notably the shift from around the Mediterranean to north-west Europe – for example the movement of Turks to Germany or Algerians to France. The European recession resulted in substantial curbs on this movement of labour to capital and from the 1970s onwards there has been a steady shift of capital to labour, usually in the form of multinational investment. Of course, it is not only cheap labour which attracts such investment but many other factors too, such as an educated workforce, good infrastructure, financial inducements and the like. The result has been a marked concentration of industrial growth in relatively few locations, such as the well-known Four Little Tigers of Pacific Asia.

Over the years, these rapidly industrialising economies have themselves become the focus for international labour migration within the Third World. Initially such movement was to the oil-rich states of the Middle East where petrodollars fuelled a huge construction and service boom in the 1970s and 1980s which was sustained largely by migrant workers from elsewhere in the region, from South Asia and also from Pacific Asia, particularly South Korea and the Philippines. The impact on the host states with their small populations has been enormous. At its peak in the late 1980s, over one-third of Saudi Arabia's labour force comprised foreign workers, over half of Libya's, while in the Gulf states the proportion was nearer to two-thirds. The Gulf War of the early 1990s, however, resulted in many being repatriated. During the 1980s, too, as oil prices fell, it was the industrial economies of Pacific Asia that became the prime target for migrant labour, usually in the unskilled areas of construction, factory work and domestic service (Figure G.1). Singapore, for example, has some 15 per cent of its labour needs met by foreign workers from Indonesia, Malaysia, Thailand and India.

Figure G.1 *Pacific Asian migration flows*

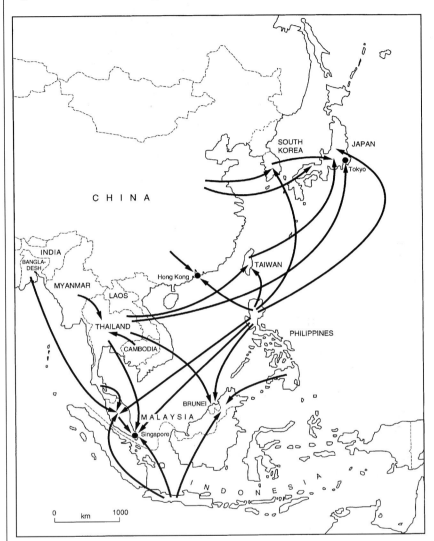

Such movements have had interesting repercussions because, as some of the sending countries began to industrialise in their turn, new labour shortages occurred as a result of labour out-migration. Thus, Thais have migrated into unskilled work in Malaysia and, in turn, Thailand has had to permit inflows of labour from Myanmar. More recently these labour flows have become infused with more skilled and professional workers whose training makes them a very attractive commodity in the regional labour market and has spurred the offer of inducements from both private firms and national governments (Jones and Findlay, 1998). But it is not only developing countries which are out to catch such migrants, over half of Australia's immigrants are from Asia and more than 70 per cent of Canada's. Wherever the migrational target, the eventual destination is usually the city and the major metropolitan centres have increasingly become more complex in ethnic terms as a consequence.

Russell King (1995) has suggested that within the complexities of the contemporary world there are five contemporary trends in international labour migration:

- the increasing globalisation of labour movements, with more countries than ever being involved as senders and hosts
- the increasing differentiation of migrants and migration – by distance, wealth and motive
- the increased acceleration and volume of migration
- the increased feminisation of overseas migrants
- the changing nature of push/pull factors, with the former becoming increasingly political or environmental as well as economic, and the latter broadening out from simple low skill/low wage industrial opportunities to encompass a variety of service and professional activities

However, we must not over-stress the role of developing countries in international migration because most movements, particularly since the fall of socialism, have occurred within Europe and North America. Moreover, the most rapidly growing form of international migration relates to refugees who now account for one-fifth of all movements. A sad indictment on our times.

Overall, however, government attempts to control and curb rural–urban migration have met with only limited success. Certainly the broadly based rural development schemes really needed for effective counter-attractions to the city seem to be beyond the means or the will of most Third World governments. The consequence in so many countries has been to introduce family planning programmes in lieu of, rather than in parallel with, development schemes. This brings the discussion to the question of natural growth.

Natural population growth and the city

In general, cities in the Third World began their recent acceleration of growth under different fertility conditions from those in the West (Gould, 1998). In contrast to nineteenth-century Europe, death rates had already begun to fall as a result of improved structural conditions, such as sanitation, medical care and nutrition. Birth rates, in contrast, are subject to a more complex behavioural set of social, economic and cultural influences and have remained relatively high, particularly amongst low-income communities wherever these are located, including cities. However, this is not an automatic or a predictable process and many demographers are emphasising the importance of national and local cultures and economies in shaping the demographic transition.

The already well-known reasons why fertility remains relatively high in the Third World, apply to the urban poor almost as much as they apply to rural areas. In particular, the need to maximise household income within the cash economy of the city ensures that most family members must work, often gleaning what returns they can from a variety of casual, part-time or informal sector jobs (see Chapter 4). Many children are, therefore, incorporated into the urban workforce. In Thailand, for example, there are as many children in the labour force as women, with over one-third being employed in factories working 10–15 hour days for half of the adult minimum wage. In addition to the immediate need for income, most families invest in children in order to provide security in old age because even the most successful industrial societies provide little in the way of welfare support. As a result, in many Third World cities, as diverse as Bogata, Delhi and Jakarta, children under 15 comprise more than half of the total population.

The arguments in favour of large families relate to individual households. In contrast, those arguments supporting family planning tend to be based on benefits of a broader societal, economic nature: smaller families lead to more domestic savings and investment, lessen pressure on limited urban resources and improve the human capital resources of the city (small families produce healthier workers). Thus reduced fertility is seen by most governments to be in the national interest. Fertility can be influenced by a variety of developmental, cultural and policy determinants, but in general these can be grouped into two categories. First are the broad socio-economic changes within society as a whole, and second are the more specific and direct attempts to change fertility at the household level. The former encompass measures which are too extensive and complex to discuss here, such as improved health care, education and employment opportunities (especially for women), all designed to alleviate poverty and redistribute wealth and thus bring about a revaluation of family priorities with regards to children. In short, these measures encourage the sort of change that has already occurred in most developed countries.

Such commitments to basic needs are expensive and long term, so many Third World governments prefer the cheaper, more direct and more immediate measures couched at the household level: for example, raising the minimum legal age for marriage, re-emphasising breast-feeding of children, but, above all, encouraging the adoption of family planning through contraception. In effect, what families in the Third World are being asked to do is to reverse the process that occurred in Europe and North America and voluntarily reduce their family size in order to bring about economic growth. Unfortunately, while the poor are expected to make maximum sacrifices in this respect, the economic benefits usually accrue to relatively few who are already wealthy and control the resources which are conserved by population control.

International agencies, perhaps fearful of the political consequences of uncontrolled population growth, have encouraged the adoption of family planning programmes. These are now officially supported by all but the most conservative governments, not only through a series of advisory and incentive schemes but also, in certain countries, by substantial disincentives to large families. In Singapore, for example, the 'Stop at Two' programme ensured that for many years families with more than two children received lower priority in housing and education waiting lists (but see Case Study H).

Although the main fertility problem still lies in the countryside, it is usually the case that family planning programmes are most effectively deployed in cities (Plate 3.3). Of course, large numbers of urban poor are characterised by high levels of fertility, for exactly the same reasons as their rural counterparts, but most of the acceptors tend to be middle-income, upwardly mobile families who would be inclined to pursue voluntary curbs to family size in any case. The urban poor, in contrast, are much more likely to 'respond', in terms of fertility, to the more comprehensive, societal changes noted above. For example, many use contraception not to limit the overall number of children they intend to have but to space them more regularly, often in line with income variations.

Plate 3.3 *Hanoi: family planning kiosk*

Case Study H

Reversing the demographic transition in Singapore

From the early years of its independence in 1965 to 1987 Singapore's national population policy was structured around fertility control. The 'Stop at Two' programme imposed a comprehensive series of fiscal and social disincentives on people to encourage them to curb their family size for the benefit of the state as well as themselves. This was very successful and the birth rate fell below replacement levels in 1975. By the 1990s, therefore, there was a clear difference in the number of children within most families, compared to the previous generation (Figure H.1).

Figure H.1 *Singapore: intergenerational differences in family size, 1990s*

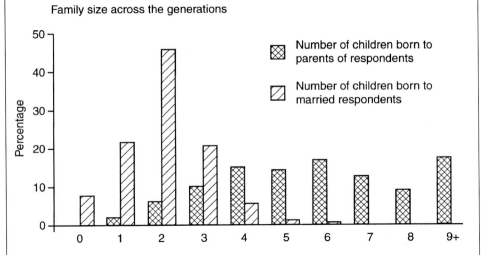

Family size across the generations

In the early 1980s, however, the Singapore government had tried to encourage higher fertility rates amongst the more educated couples, largely because it believed that the labour demands of the new high-tech industries obliged Singapore to explore to the full its gene pool. A Social Development Unit was established to bring together university graduates and other well-qualified individuals in order to encourage marriage. In addition large tax incentives were offered to such couples to encourage them to have three or more children. In contrast, the less intelligent were offered financial inducements to undergo sterilisation after only one child (Drakakis-Smith and Graham, 1996).

By the late 1980s, however, it became clear that the absolute size of the labour force would begin to diminish within about twenty years and so a new population policy was introduced which encouraged *all* families to have 'three or more, if you can afford it', in an attempt to reverse the demographic transition (Plate H.1). However, despite a range of incentives being made more widely available, the response has been limited, largely due to a resistance by more affluent families to being told how to organise their life, together with a growing lack of enthusiasm on the part of young Singaporeans for the rather authoritarian attitude

Plate H.1 *Singapore: family planning poster*

Figure H.2 *Singapore: fertility rates before and after the introduction of the new population policy*

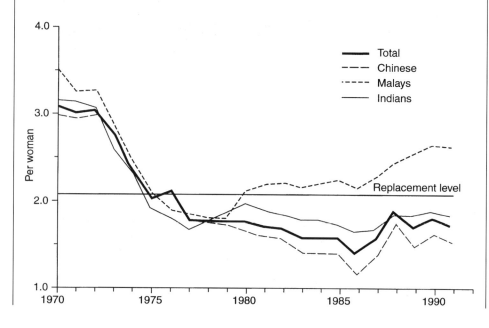

of the government. As a result the vital population indicators are all continuing to move in the wrong direction (Figure H.2) (Graham, 1995). Between the mid-1980s and mid-1990s the mean age at first marriage rose from 27.9 to 29.4 and from 24.9 to 26.3 for men and women respectively, while the age of mother at first birth has risen from 26.3 to 27.8. Clearly, reversing the demographic transition in order to meet labour needs has proved to be a difficult if not impossible task. However, the emphasis on the family and family life has continued to play a major role in government social policy, largely because it is seen as an appropriate Asian response in the face of increasing welfare needs, particularly for the elderly (Plate H.2).

Plate H.2 *Singapore: family values in care of the elderly*

Clearly, urbanisation and demographic change are strongly linked, but not just in the traditional ways assumed for so many years. Complex patterns of fertility exist even within the same city for different social, economic or cultural groups. In the more advanced economies of the Third World, for example, concerns are even growing about the ageing of the workforce and the consequent impact this will have, not just on economic production but also in terms of social welfare. Few public schemes exist to support the aged, and the role of the family in accepting responsibility is being heavily emphasised in countries such as Singapore and Malaysia. Meanwhile, in other parts of Asia and Africa, change in urban fertility continues to occur almost without reference to economic growth.

It is at this point that such complex arguments about urban population growth turn full circle. High rates of growth, whether natural or migrational, are usually the consequence of development disparities, both social and spatial, not the cause. Yet, over the last decade, overseas aid for general development has fallen, while funds specifically tied to the expansion of family planning programmes have not. As a result the quality of life for many of the urban poor has continued to deteriorate.

Key ideas

1. Migrational and natural growth rates vary in relative importance in different cities.
2. Rural–urban migration has changed in character in recent years.
3. Government responses have largely been ineffective in preventing rural–urban migration.
4. High fertility persists in many cities, particularly amongst poor families.
5 International labour migration is slowly expanding and introducing a new dimension to urban demographic change.

Discussion questions

* Assess the relative importance of the migrational and natural components of urban growth in the Third World.
* Is there a typical migrant?
* Are fertility control programmes the answer to rapid urban population growth?
* What are the benefits and costs of rapid urban population growth to municipal government?

References and further reading

Baker, J. and Pedersen, P.O. (1992) *The Rural–Urban Interface in Africa*, Nordic Institute for African Studies, Uppsala.

Dixon, C. (1990) *Rural Development in the Third World*, Routledge, London.

Drakakis-Smith, D. and Dixon, C. (1997) 'Sustainable urbanisation in Vietnam', *Geoforum*, 28(1): 21–38.

Drakakis-Smith, D. and Graham, E. (1996) 'Shaping the nation-state: ethnicity, class and the new population policy in Singapore', *International Journal of Population Studies*, 2(1): 69–90.

Findlay, A.M. (1994) *The Arab World*, Routledge, London.

Gould, W. (1998) 'African mortality and the new urban penalty', *Health and Place*, 4(2): 171–181.

Gould, W.T.S. and Findlay, A.M. (eds) (1994) *Population, Migration and the Changing World Order*, Wiley, London.

Graham, E. (1995) 'Singapore in the 1990s: can population policies reverse the demographic transition', *Applied Geography*, 15: 219–232.

Grant, M. (1995) 'Movement patterns and the medium-sized city', *Habitat International*, 19: 357–370.

Jones, H. and Findlay, A.M. (1998) 'Regional economic integration and the emergence of the East Asian international migration system', *Geoforum*, 20(1): 87–104.

King, R.C. (1995) 'Migration, globalization and place', in D. Massey and P. Jess (eds) *A Place in the World*, OUP, Oxford: 5–44.

Lockwood, M.C. (1995) 'Population and environmental change: the case of Africa', in P. Sarre and J. Blunden (eds) *An Overcrowded World*, OUP, Oxford: 59–108.

Parnwell, M. (1993) *Population Movements and the Third World*, Routledge, London.

Population Concern (1994) *Fact File*, Population Concern, London.

Rigg, J.C. (1996) *Southeast Asia*, Routledge, London.

Sage, C. (1996) 'The search for sustainable livelihoods in Indonesian transmigrant settlements', in R. Bryant and M. Parnwell (eds) *Environmental Change in Southeast Asia*, Routledge, London: 97–122.

Simon, D. (1996) *Transport and Development in the Third World*, Routledge, London.

Todaro, M.P. (1994) *Economic Development*, Longman, New York.

UNFPA (1993) *The State of World Population*, United Nations, New York.

⟨4⟩ Urban environmental matters

Introduction: identifying the brown agenda

For many years cities have been seen as the villains of the world's increasing environmental problems, consuming scarce resources from, and spewing waste into, their surrounding regions. Only slowly has an awareness of these cities' own environmental problems emerged as worthy of investigation in their own right. While international gatherings, such as the Earth Summit of 1992, have helped to develop this awareness and have encouraged more detailed documentation of urban environmental matters, it is the increased incidence of major disasters occurring within cities which has focused attention more sharply. Examples from the 1990s would include the return of cholera to many Latin American cities, the incidence of plague in Surat in India, and the Cairo earthquake.

The nature of these urban environmental matters is wide ranging, involving such problems as pollution, land degradation or hazardous living and working conditions. Collectively termed the 'brown agenda', such problems vary immensely according to a combination of contributing factors. First, there is the nature of the urbanisation process itself – the rate and scale of growth, together with the degree of concentration of the population. Even quite small settlements can produce intense development problems. For example, Jenny Bryant-Tokelau (1994) has written of Ebeye in the Marshall Islands of the South Pacific. This town has a population of only 9,000 but its atoll location concentrates these into densities of 23,000 per square

Plate 4.1 *Hong Kong: housing collapse after heavy rain*

kilometre and results in severe problems of overcrowding and waste disposal.

Also important are the characteristics of the natural ecosystems within which the settlement is located (Plate 4.1). Many cities are vulnerable to natural hazards, such as seismic activity, or else are located in environmentally sensitive areas on coastlines or in arid regions which affect both inputs, such as water resources, and outputs, such as solid waste disposal. Climatic conditions too can influence the brown agenda, such as the air pollution in Mexico City (Case Study I).

The third factor is the level and nature of development itself. This can affect both the pressures felt by urban residents (Figure 4.1) and the ability of the household and the authorities to respond to these pressures. In general, the latter improves with increased incomes (Figure 4.2) but this relationship also depends on the motivation and capacity of local government to recognise and respond to increasingly complex brown agendas. In general, there is a time lag between economic growth and environmental policy development and enforcement; but this period can vary considerably between cities at the same developmental level. For example, Mexico City and Kuala Lumpur are capital cities of countries with similar per capita GDP, but the response to urban environmental problems has been much stronger in the former than the latter, largely resultant from the relative success of NGO lobbying. In general, it is difficult not to agree

Case Study I

Air pollution in Mexico City

Mexico City Metropolitan Area is one of the largest and most polluted cities in the world. Its current population stands at 17 million people and this is predicted to rise to 25 million by 2010. Ozone concentrations are the highest in the world, and suspended particulate levels amongst the highest. Carbon monoxide and lead pollution are well above advisory levels. These air pollutants account for a huge number of deaths, illnesses and, most important of all for many authorities, lost workdays. It is estimated that fine particulate matter and ozone together account for over 20 million lost workdays each year, largely through respiratory illnesses. Altogether the total economic loss due to the health effects of air pollution is conservatively estimated at US$1.5 billion per year.

In 1990 the authorities adopted an integrated policy against air pollution, which included measures such as a one day per week car ban and vehicle inspections. However, little change in the overall level of air pollution occurred, largely because it proved difficult to change social habits – many people bought another car, usually old, rather than stay off the roads for one day.

New measures have been based more on fiscal and technological change, rather than behavioural, and have been prioritised by cost effectiveness. Taxes on petrol, for example, raise revenues as well as cutting car use. However, effective policy responses, even in the single field of transport planning, are hampered by the fact that the Mexico City Metropolitan Area covers 17 municipalities, together with the Federal District itself. Creating common policy responses will prove crucial to the success of any externally prompted initiatives, such as those from the World Bank.

Source: Carl Bartone *et al.* (1994).

with Ernesto Pernia (1992) who believes that not only do most Third World governments lack sufficient knowledge and information on environmental matters, but also see environmental issues as secondary to the more pressing problems of economic growth and to politics. Many resent pro-environmental 'interference' from the West, with Malaysian Prime Minister, Mohammed Mahathir and others interpreting this as an attempt to reduce the competitive advantages of developing countries!

In general, the environmental pressures of the brown agenda come from two different directions (Figure 4.3). The first is related to the development priorities given to economic growth, often without thought or action related to the environmental (and social)

Figure 4.1 *Environmental indicators and GNP per capita*

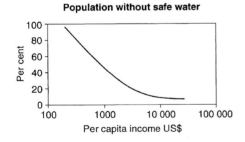

Population without safe water

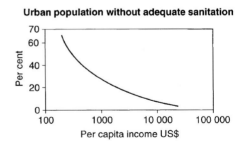

Urban population without adequate sanitation

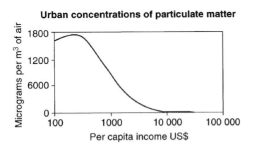

Urban concentrations of particulate matter

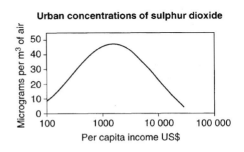

Urban concentrations of sulphur dioxide

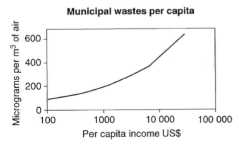

Municipal wastes per capita

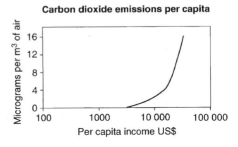

Carbon dioxide emissions per capita

Source: World Bank, *World Development Report* (1992).

consequences. In too many countries the philosophy is 'develop now and clean up later'. Despite the arguments of environmental economists that market forces will bring about change as a result of the costs associated with irresponsible production, industrialists have responded poorly unless coerced by the enactment and enforcement of appropriate legislation. This is because many of the consequences of industrial pollution, such as lead poisoning, are not recognised by factory owners as having direct production costs. However, it is not only industrialists who contribute to environmental problems through the operations of the market economy. The built environment itself offers many opportunities for enormous profits through land deals,

Figure 4.2 *National prosperity and protection against environmental health hazards*

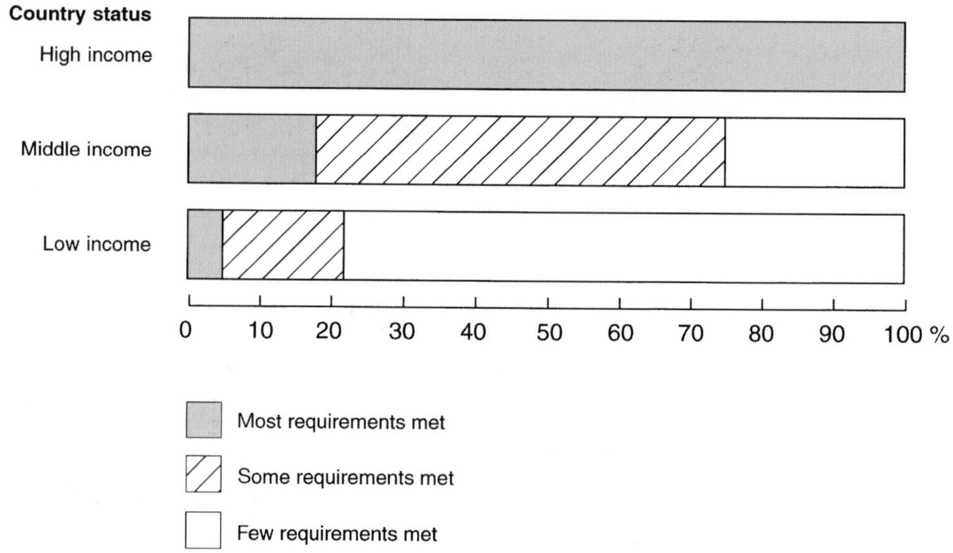

Source: After Bartone *et al.* (1994: 18).

construction projects or the provision of public utilities – but only to those who can afford to pay. Not only has the circuit of capital investment in the built environment been largely controlled by the planning process but urban managers themselves have direct involvement in urban development projects and are frequent beneficiaries of uncontrolled development.

The second major contributor to the brown agenda is poverty. Rapid population growth rates, continuing high birth rates and insufficient formal sector employment, particularly under structural adjustment, have all helped to increase urban poverty substantially. Not only are many families living below the poverty line but many more are vulnerable to the sudden changes in fortune that can affect low-income households, such as illness or eviction, and push marginal families over the edge and into poverty. It is difficult to assess the extent of poverty and vulnerability in Third World cities, as related statistical information is very unreliable. Undoubtedly, however, urban poverty is growing rapidly and in some countries the proportion of poverty in cities is greater than in rural areas (Table 4.1). In most cities the average income is often distorted by a relatively small number of wealthy households, concealing the widespread incidence of poverty and vulnerability. In Vietnam, for example, the average incomes for Hanoi and Ho Chi Minh City are 40 per cent and

Figure 4.3 *Main sources of environmental pressure in Third World cities*

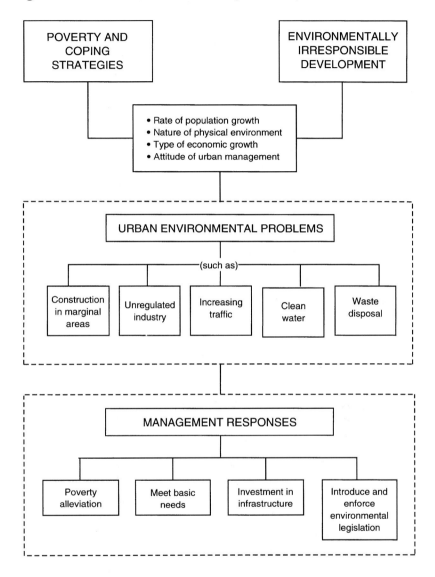

170 per cent higher than the national average, but the income distribution within these and other cities is so skewed that fully one-third of the urban population of Vietnam has incomes lower than the rural average. Moreover, rural dwellers are more likely to have savings (however small) and to have alternative food sources available to them (Drakakis-Smith and Dixon 1997).

The consequence of so much poverty is a large and increasing number of urban residents who simply seek to survive as best they can,

Table 4.1 *Absolute poverty in urban and rural areas (percentage below poverty line)*

	Urban areas	Rural areas	Percentage urban	Ratio of rural poor to urban poor*
Africa				
Botswana	30.0	64.0	29	4.8
Côte d'Ivoire	30.0	26.0	41	1.3
Egypt	34.0	33.7	47	1.1
Morocco	28.0	32.0	49	1.2
Mozambique	40.0	70.0	28	4.5
Tunisia	7.3	5.7	55	0.7
Uganda	25.0	33.0	11	10.3
Asia				
Bangladesh	58.2	41.3	17	3.4
China	0.4	11.5	60	18.9
India	37.1	38.7	27	2.8
Indonesia	20.1	16.4	31	1.8
South Korea	4.6	4.4	73	0.3
Malaysia	8.3	22.4	44	3.2
Nepal	19.2	43.1	10	21.0
Pakistan	25.0	31.0	33	2.5
Philippines	40.0	54.1	43	1.8
Sri Lanka	27.6	45.7	22	6.1
Latin America				
Argentina	14.6	19.7	87	0.2
Brazil	37.7	65.9	76	0.6
Colombia	44.5	40.2	71	0.4
Costa Rica	11.6	32.7	48	3.2
Guatemala	61.4	85.4	40	2.1
Haiti	65.0	80.0	29	3.1
Honduras	73.0	80.2	45	1.3
Mexico	30.2	50.5	73	0.6
Panama	29.7	51.9	54	1.3
Peru	44.5	63.8	71	0.6
Uruguay	19.3	28.7	86	0.2
Venezuela	24.8	42.2	85	0.3

Sources: UNCHS (1996), World Bank, *World Development Report* (1994).

* Note: In absolute totals: less than 1.0 indicates more urban poor than rural poor.

seemingly oblivious of the cost to the environment. This does not mean that the poor are unaware of environmental matters or of the consequences of their actions. It is just that they are too poor to be able to do much about it. Any improvement in their situation and the consequent easing of pressures on the urban environment is, therefore, going to be wholly dependent on either alleviating their poverty or on state action to meet their basic needs – preferably both.

The combination of poverty and unregulated economic growth gives rise to a range of urban environmental problems (Bartone *et al.* 1994) and it is important to realise that they extend over a wide spatial scale – from the household and its immediate environs to the regional and even global scale (Table 4.2). In general, the concerns at the household, workplace or community levels are more immediate and

Table 4.2 *Spatial dimensions of the brown agenda*

	Principal service infrastructure	Problem issues
Household/workplace	Shelter	Substandard housing
	Water provision	Lack of water, expensive
	Toilets	No sanitation
	Solid waste	No storage
	Ventilation	Air pollution
Community	Piped water	Inadequate reticulation
	Sewerage system	Human waste pollution
	Drainage	Flooding
	Waste collection	Dumping
	Streets (safety)	Congestion, noise
City	Industry	Accidents, hazards, air pollution
	Transport	Congestion, noise, air pollution
	Waste treatment	Inadequate, seepage
	Landfill	Unmonitored, toxic, seepage
	Energy	Unequal access
	Geomorphology	Natural hazards
Region	Ecology	Pollution, deforestation, degradation
	Water sources	Pollution, overuse
	Energy sources	Overextended, pollution

Source: Bartone *et al.* (1994: 15).

relate primarily to health and equity in access to basic services. Those at the regional and global levels are more long term in nature and are linked to the impact of resource use and depletion on future generations. Between these two is the city itself, combining all these issues in a complex situation that requires very careful management in order to bring about sustainable urbanisation for the benefit of all. The remaining sections of this chapter will address the main environmental problems of Third World cities in the context of their scale and immediacy, before proceeding to an examination of the management implications.

Environmental problems at the household and community level

Less than a hundred years ago the death rates in European and North American cities from diseases such as cholera, tuberculosis or typhoid were as high as they are in many Third World cities today. What is most welcome about the changes which medical, hygiene and planning improvements have brought about is the way in which low-income groups have also benefited. Admittedly their health and life expectancy in the West are still inferior to those of wealthier groups, largely due to their immediate living and working environments, but the gap is much closer than that which currently exists in most developing countries, where it is estimated (UNCHS, 1996) that some 600 million urban residents live in circumstances that continually threaten their health (Table 4.3).

For most of the urban poor, food is their main priority, and this can often take up the bulk of their meagre cash income (Plate 4.2). Shelter needs are met as best they can, given the minimal resources available, often in squatter shacks or tenement rooms that not only provide few of the facilities necessary to sustain a healthy life but actually contribute to ill-health, more so if the accommodation is located in an unhealthy neighbourhood environment. Table 4.3 lists the principal features of the housing environment that impact on the household. Four of these – water, sewerage, air pollution and overcrowding – are focused on in the most recent UN Habitat Report (UNCHS, 1996) as being particularly problematic.

Of all the basic needs, access to clean water is probably the most important (Plate 4.3). The World Bank estimates that at least 170 million urban residents in developing countries lack access to potable water *near* (not in) their homes. In Jakarta, for example, a city of almost ten million people, only 40 per cent have access to safe

Table 4.3   Aspects of the housing environment and health

The structure of the shelter (which includes a consideration of the extent to which the shelter protects the occupants from extremes of heat or cold, insulation against noise and invasion by dust, rain, insects and rodents).

The extent to which the provision for water supplied is adequate – both from a qualitative and quantitative point of view.

The effectiveness of provision for the disposal (and subsequent management) of excreta and liquid and solid wastes.

The quality of housing site including household accidents and airborne infections whose transmission is increased: acute respiratory infectious diseases, pneumonia, tuberculosis.

The presence of indoor air pollution associated with fuels and used for cooking and/or heating.

Food safety standard – including the extent to which the shelter has adequate provision for storing food to protect it against spoilage and contamination.

Vectors and hosts of disease associated with the domestic and peri-domestic environment.

The home as a workplace – where occupational health questions such as the use and storage of toxic or hazardous chemicals and health and safety aspects of equipment used need consideration.

Source: UNCHS (1996).

drinking water; for urban Indonesia as a whole the proportion is down to one-third, indicating an even worse situation in secondary cities. Moreover, those who have access to the water tend to be the better-off who live within the reticulation system or who can afford bore wells. The marginal poor, those whose health is most at risk, have worst access. In Lima, for example, only 7 per cent of houses in the peripheral areas have access to safe drinking water. Families who do not have direct access must buy their water from commercial vendors at prices far in excess of those charged for piped water (Table 4.4). Thus the urban poor must pay more than the wealthy for clean water – little wonder that they must sometimes resort to using contaminated water when money is not available, with disastrous consequences for their health in the form of intestinal parasites, typhoid and diarrhoeal diseases (Figure 4.4). In Brazil it is reported that those without access to piped water are almost five times more likely to die from diarrhoeal diseases; in Jakarta they are responsible for 20 per cent of deaths for children aged five or less (World Resources Institute, 1996). Those who do not die are often severely debilitated by disease and/or parasites, affecting their ability to work and earn enough to improve the living environment.

Plate 4.2 *Suva market: food needs must be met in the cash economy in cities; those who cannot pay are vulnerable to nutritional problems*

The problems of water supply have forced many urban authorities to over-exploit fragile sources, such as underground aquifers, and there are many examples of falling water levels in cities such as Manila and Jakarta. In some coastal cities, such as Lima and Dakar, this has drawn in saline intrusions. In other cities, such as Mexico City and Hanoi, aquifer depletion has resulted in widespread subsidence and frequent flooding during rainy seasons. In Bangkok parts of the city have sunk by more than 1.5 metres since the 1930s. Ironically, one of the major causes of water shortage and pollution is the construction of golf courses. In Thailand alone some 200 have been built over the last ten years, each of which uses between 3 and 6 million litres of water per day. In addition tonnes of chemical fertilisers and pesticides are also applied, eventually seeping through into the aquifers and streams (see Elliott, 1999).

Closely linked to water provision is the removal of human waste through sewerage systems (Case Study J). It is estimated that during the 1980s the number of urban residents in the Third World without access to adequate sanitation increased by almost 25 per cent to 400 million. Even where such systems exist, waste is often dumped untreated into rivers or seas with disastrous effects on aquatic life. Lack of adequate toilet facilities also pollutes groundwater, as well as directly contributing to the proliferation of those vectors that breed in such conditions. The health problems created by poor water and sanitation facilities are compounded by overcrowding and poor ventilation which intensifies the transfer of respiratory infections. Tuberculosis is still the major killer amongst adults in developing countries, while pneumonia or bronchitis is 50 times more likely to lead to death amongst Third World children than in more developed

Plate 4.3 *Access to clean water is vital for improved health. Water standpipe in Epworth squatter settlement, Harare*

Table 4.4 *Water charges in selected cities*

	Average piped tariff US$ M^3	Private vendor tariff US$ M^3	Radio private/ public
Jakarta	0.363	1.848	5.1
Bandung	0.268	6.161	23.0
Manila	0.232	1.873	8.1
Calcutta	0.049	2.099	42.8
Madras	0.046	0.875	19.0
Karachi	0.047	1.747	37.2
Ho Chi Minh City	0.045	1.511	33.6

Source: Asian Development Bank (1993).

countries. The impact of overcrowding and poor ventilation is also exacerbated by indoor pollution, largely through the use of biomass fuel for heating and/or cooking. Bronchitis tends, therefore, to be more prevalent amongst those most frequently involved in these domestic activities – women and children.

Figure 4.4 *Water pollution and its health consequences*

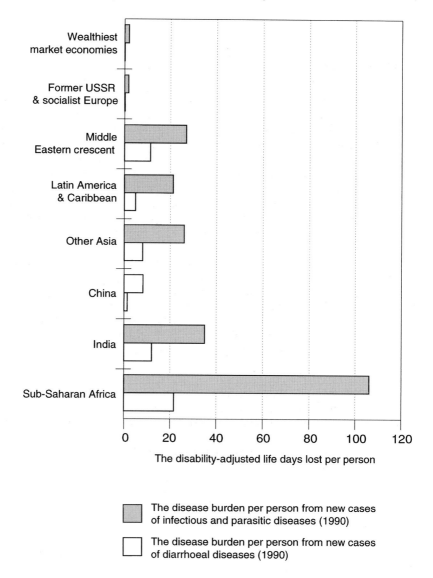

The disability-adjusted life days lost per person

[gray box] The disease burden per person from new cases of infectious and parasitic diseases (1990)

[white box] The disease burden per person from new cases of diarrhoeal diseases (1990)

The city environment

Within the city as a whole, a variety of circumstances combine to threaten sustainability both directly and indirectly, usually through various forms of pollution. This section will first describe the direct threats and then address the consequential pollution effects.

One of the main consequences of unregulated and unplanned urban development is an increasing expansion into physically marginal

Case Study J

Water, sewage and waste in Changzhou

Changzhou lies in the Yangtze River delta about 160 kilometres from Shanghai. Like many Chinese cities its official boundaries cover a wide area of some 190 square kilometres and within this there is an urban population of some 700,000, although together with three nearby cities the overall urban population is in excess of three million (Figure J.1). Since the establishment of the People's Republic of China, Changzhou has developed its industrial base rapidly, mainly textiles, mechanical engineering and chemicals. Since 1980 the average growth rate in industrial output has been around 20 per cent per year. Such economic success has attracted much population migration and has placed enormous burdens on the urban infrastructure, which has not been able to keep pace with overall growth, so that only 21 per cent of discharged sewage, for example, can be treated.

Although Changzhou is located on the alluvial plain of Taihu Lake and has generous surface and aquifer supplies of water, the city suffers from a serious water shortage. This is largely due to the decreasing aquifer resources which are largely being depleted (and polluted) by industrial demands. Not only are public supplies diminishing but private wells are beginning to dry up, or have been abandoned because of pollution. Surface water supplies too have been badly affected by industrial discharge, and the main waterway that runs through the city, the Beijing–Hangzhou Grand Canal, can no longer be used for irrigation or aquaculture.

The response to the crisis has largely been reactive rather than preventative, with water being brought in from the Yangtze River, 25 kilometres away. There have also been attempts to increase public awareness (and prices) in order to bring about savings in domestic

Figure J.1 *Changzhou: population growth and cultivated land*

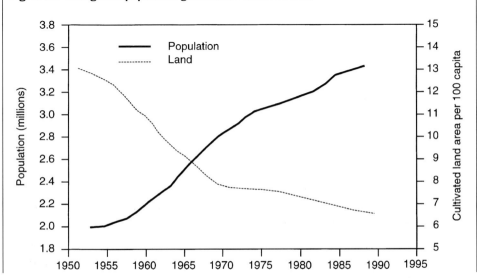

consumption. Actions involving industrial plant, either to encourage recycling of water or to curb discharges, receive relatively limited attention. Sustained economic growth is clearly considered to be more important than sustainable urbanisation.

Currently garbage production is growing at about 5 per cent per year, faster than overall population growth. At present rubbish is mostly collected and dumped at landfill sites where it is eating into agricultural land at the rate of 5 hectares per year. At present the treatment of collected solid waste is very limited and the urban authorities realise that more treatment capacity is desperately needed. In line with current advice, they are seeking to raise the necessary funds by charging fees for collection and disposal. It remains to be seen whether low income households will comply with this change, or will withdraw from public services and create an even greater environmental problem.

Source: various reports in *Ambio*, 25(2), 1996.

areas. A high proportion of the major cities of the Third World are located in hazard-vulnerable zones anyway, but the impact of natural disasters is often compounded by inadequate planning safeguards, inappropriate building design and the sheer overcrowding that forces people into physically marginal areas (see Plate 4.1). Major catastrophes are intensified in their impact by these factors; for example the collapse of many buildings through poor construction in the Cairo earthquake of 1994. In 1988 the huge floods and landslides in Rio de Janeiro were caused not only by exceptional rains and uncontrolled favala construction but also by neglected, blocked and inadequate drainage systems.

Natural hazards are often accompanied by man-made problems, many of which relate to inadequately controlled industrial development. Most Third World countries are so desperate for industrial investment that they are prepared to waive the few regulatory controls that exist. For example, the Philippines government advertised in *Fortune* magazine that 'to attract companies like yours . . . we have felled mountains, razed jungles, filled swamps, moved rivers, relocated towns and in their place built power plants, dams and roads'. Once in place, activities are often unregulated in both small and large concerns and industrial accidents claim thousands of lives each year. In addition, air and water pollution from untreated or uncontrolled industrial emissions and wastes exacerbate already poor situations resultant from domestic pollutions. Perhaps the most infamous example of the former occurred in Bhopal, where the Union Carbide gas escape in 1984 caused 3,300 deaths and 150,000 serious injuries.

It must be noted at this point that in the Third World as a whole most urban industrial pollution originates from small-scale enterprises,

Case Study K

Tanneries in Leon, Mexico

Of the tanneries in Mexico, just over-one-third are classified as small-scale. Geographically, two-thirds of these are located in the city of Leon, where 58 per cent of the leather and 54 per cent of the shoes in Mexico are produced. In Leon, three-quarters of the tanneries are cottage enterprises with 1–6 employees, one-fifth are considered to be small-scale with 7–16 employees, and the remainder are considered medium-scale with 17 or more employees (Figure K.1). Almost all the leather produced in Leon originates from salted hides which are tanned with chromium. Cottage tanneries each process on average 400 hides per month, the average produced by each small-scale enterprise is about 2,500 hides per month. Wastes generated in the tanneries include large volumes of effluents containing salt, hair, sulphide, chromium and other additives which contribute to the total dissolved solids (TDS). Solids include split, fleshings and shavings (Figure K.2).

Figure K.1 *Leon, Mexico: tannery size*

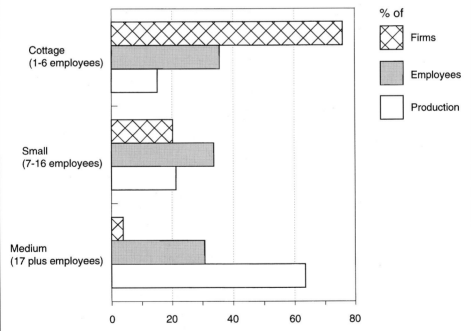

None of the tanneries in the city of Leon have wastewater treatment systems. Those tanneries discharging to the sewer system, however, are required by the city water and sewage utility to install solids traps to prevent clogging of the sewers. These traps have been installed in most, but not all, registered connections. Although the Tanneries Association claim that the traps are cleaned two to three times a year, data from the water and sewage utility, which provides the sludge removal services, indicate that less than two-thirds of them are adequately serviced. Furthermore, the solids collected from the traps are dumped without treatment directly into the Turbio River downstream from the city.

Figure K.2 *Leon, Mexico: waste production*

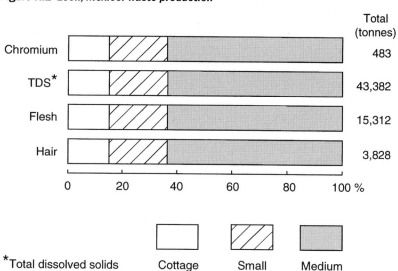

The main environmental problem caused by the operation of the tanning industry in Leon is the high TDS content of the effluents, which preclude their reuse. Other problems include the salt contamination of wells located south of the city and the increase in the chromium content of these wells. The main solid waste of the industry is fleshings, which are used for the production of tallow in makeshift factories located some 15 kilometres from the city. The transport of the solids is done in open trucks that spill both liquid and solids into the streets of the city. The fleshings are deposited on the ground, and the liquids leaching from the piles run into surface water courses. The tallow-making process generates noxious gases, produced by the reaction of sulphide with the acid, and from decomposition of amino acids. Liquid wastes from the process are disposed in impoundments where anaerobic decomposition occurs, causing bad odours in the neighbourhood. Other solid waste, such as hair, is disposed together with the wastewaters.

The Center for Tanneries Research and Technical Assistance of the State of Guanajuato (CIATEG) demonstrated that the dehairing and tanning liquids could be recycled, but was not successful in achieving implementation of these process modifications. Failure was due to lack of technical capacity, shortage of financial resources, and physical space limitations in the SSI tanneries. Mexican legislation has established standards for wastewater discharged to sewers and water bodies. In addition, hazardous waste management legislation has been enacted whereby any wastes containing more than 5 mg[-1] of total chromium must be disposed of in specially designed disposal sites. The city of Leon, conscious of these environmental problems, has initiated a programme to relocate many of its tanneries to an industrial estate, where treatment systems for the effluents will be provided. The tanners will have five years to relocate their unit processes up to and including the tanning itself. The treatment plant will be financed with grants from state and federal funds, covering 70 per cent of the total cost. The remainder will be financed by the tanners and by the local water and sewage utility.

Source: Bartone *et al.* (1994).

particularly those related to food processing, leather tanning, textile dyeing, various forms of metalworking, such as electroplating, and engineering (Case Study K). The reasons why small-scale industries contribute disproportionately are threefold. First, control measures tend to be directed to larger firms, and foreign companies in particular feel some need to comply. Second, control measures are often more expensive per production unit and beyond the means of most small firms. Third, small enterprises tend to use older or less efficient equipment that is more prone to pollute. Many small firms are, of course, illegal or semi-legal in the way they operate, so there is little incentive for them to go clean, in all senses of the word. Monitoring is expensive and seldom undertaken by the authorities who, if making any gesture towards pollution control, are more concerned with the public relations coup that action against large-scale firms will produce.

Air pollution is of course affected by more than just industrial emissions; important as these are, extensive electricity production from, and domestic use of, fossil fuels, together with increasing vehicle ownership have both intensified the situation. Vehicle ownership within Third World cities is increasing at between 5 and 10 per cent per annum, doubling every seven years in some Asian cities, usually with vehicles that are old, inefficient, noisy and emit a range of pollutants. The result is extensive contamination in most Third World cities, but particularly in the largest where traffic, industry and domestic energy use is most extensive and concentrated (Table 4.5). In Bangkok 26 million workdays per year are lost through respiratory problems. In Beijing (Plate 4.4) the situation is so bad that British Embassy staff have been instructed to reside above the eighth floor of their residential blocks in order to avoid 'Beijing throat' (Cook, 1993). It must be noted at this point that Third World cities are not alone in their level of pollution. In Katowice in Poland air pollution is so bad that male life expectancy has fallen over the last two decades and more than one-third of the children suffer from bronchitis. It is also salutary to remember that the bulk of global carbon emissions still come from the West and that the three leading offenders on a per capita basis are the USA, Canada and Australia.

Water pollution too, as noted earlier, is also extensive in and around Third World cities. In most instances this is the consequence of inadequate sewage and sanitation facilities, compounded by poor solid waste disposal services (Plate 4.5). As economic growth increases so does the amount of waste generated per person; so in Hong Kong and Singapore daily waste generation is about 86 kilos per person,

Table 4.5 *Overview of air quality in selected large cities*

City	Sulphur dioxide	Suspended partic.	Airborne lead	Carbon monoxide	Nitrogen dioxide	Ozone
Bangkok	Low	Serious	Above guideline	Low	Low	Low
Beijing (Peking)	Serious	Serious	Low	(no data)	Low	Above guideline
Bombay	Low	Serious	Low	Low	Low	—
Buenos Aires	—	Above guideline	Low	—	—	—
Cairo	—	Serious	Serious	Above guideline	—	—
Calcutta	Low	Serious	Low		Low	—
Delhi	Low	Serious	Low	Low	Low	—
Jakarta	Low	Serious	Above guideline	Above guideline	Low	Above guideline
Karachi	Low	Serious	Serious		(no data)	
Manila	Low	Serious	Above guideline		(no data)	
Mexico City	Serious	Serious	Above guideline	Serious	Above guideline	Serious
Rio de Janeiro	Above guideline	Above guideline	Low	Low	(no data)	—
São Paulo	Low	Above guideline	—	—	—	—
Seoul	Serious	Serious	Low	Low	Low	Low
Shanghai	Above guideline	Serious	—	—	—	—

Serious:	WHO guidelines exceeded by more than a factor of 2.
Above guidelines:	WHO guidelines exceeded by up to a factor of 2 (short-term guidelines exceeded on a regular basis at certain locations).
Low:	WHO guidelines normally met (short-term guidelines may be exceeded occasionally).
—	Insufficient data

Source: UNCHS (1996).

Plate 4.4 *Beijing: a typical smoggy morning in the city centre*

Plate 4.5 *Hanoi: household rubbish collection in the ancient quarter*

compared to nearer 50 kilos for Indonesian and Filipino cities. Unfortunately, economic growth does not necessarily correlate with adequate waste disposal facilities and both Taiwan and South Korea are far behind in this respect. Even where a high proportion of domestic waste is collected, most is simply dumped in open sites or else incinerated without proper control, adding to the air pollution problem. Much industrial waste, of course, can be hazardous in its own right and there are many instances of deaths directly related to such waste. Others suffer, ironically, from trying to recycle waste and contracting a wide-range of diseases (Case Study L). In Manila, some 20,000 people live on the garbage dump known as Smokey Mountain; they collect recyclable material in order to make a living while their children play in the rubbish.

Increasingly, many of the cities in the Third World, particularly those in the more industrialised states, are being obliged to adopt more stringent environmental legislation, especially to control industrial pollution. However, much of this legislation is unenforced or applied only to new plant, not to existing factories. Others have simply exported the dirty components of their manufacturing industry to neighbouring states grateful for any kind of investment – China and Mexico, in particular, have reputations for welcoming such dubious gifts.

The urban footprint

William Rees (1992) has developed the 'ecological footprint' concept to evaluate the impact of the city on its surrounding area. This is based on the old concept of the 'carrying capacity' of an ecological region – the maximum rate of resource consumption and waste discharge that can be sustained over an indefinite period without impairing the integrity of the system itself. Clearly cities reach beyond their immediate area and draw on the resources of a wide and variable set of hinterlands, thus setting up situations of regional deficit and surplus as resources (and waste) are shifted between regions. Such ecological deficits are not necessarily problematic if the net carrying capacity across and between ecological regions remains in balance. However, as Rees notes, economic pressures and trends are usually ignorant of carrying capacities and related sustainabilities. It is the net negativities of this relationship that constitute the urban footprint.

Of course, these footprints, like those on a sandy beach at holiday time, can vary in size, depth and degree of overlap, often being difficult to distinguish as separate entities. For example, the fact that many cities draw their food from a wide range of sources can often mean that agricultural zones close to the city are relinquished for expansion of the built environment. On the other hand, growing

Case Study L

Scavengers in Calcutta

The populations of the Tikiapara and Tiljala districts of Calcutta amount to around 110,000 people. Most are in low earning activities, many live in squatter huts (bustees), and some 1,300 people earn a living by picking over street garbage to salvage recoverable and sellable materials. Most pickers are males, many are children and few have any formal education.

In cities such as Calcutta most of the recyclable material is saved for resale by households or shops. Much of the rest is collected by itinerant waste collectors. It is what is left that is scavenged by the street pickers – mostly paper, plastics and glass with occasional pieces of metal. Little processing occurs before the material is passed on to small waste dealers. Most of the pickers earn below 600 rupees (£12) per month from this work which exposes them to many health risks, particularly as they can afford no protective measures.

Most pickers have a regular area in which they work, and take their materials to the same waste recycler. They have little idea what happens to it after they receive their money, having many other priorities with which they concern themselves. The pickers recognise their low status in society – one which is intensified by suspicions that they also steal. Thus their low earnings are matched by low status and low self-esteem.

The life of street scavengers appears to be different from scavengers who work on municipal dumps in Calcutta and other major cities. The latter seem to be more organised, partly because they need to pay fees to the dump managers or even obtain licences, as in Harare (Tevera, 1993). In contrast street pickers lack self-motivation and self-belief. However, as the general public becomes more aware of the role of waste recycling, and as the scavengers themselves appreciate the benefits of organisation, there may be some small improvements. In Calcutta the United Bustee Development Association has begun to form a pickers' organisation and to offer education for them and their children. Given the fact that in many cities like Calcutta more formal waste collection and recycling services are unlikely to materialise for many years, this positive incorporation into the community of scavengers and scavenging must be welcomed.

Source: this is a summary of Christine Furedy and Mohammed Alamgis (1992) 'Street pickers in Calcutta Slums', *Environment and Urbanisation*, 4(2): 54–59.

poverty in African cities has meant that open spaces in and around the city are increasingly being used for subsistence food cultivation.

To clarify the nature of the urban footprint, it is useful to simplify the spatial dimensions of the discussion into two parts: first, the immediate peri-urban areas where the footprint looms large and heavy; and second, the broader regional resource/waste impacts. With regard to the former, Habitat has identified three areas of particular concern (UNCHS, 1996):

- Unplanned and uncontrolled urban sprawl – often through the expansion of illegal squatter settlements located on marginally usable land. However, there are many other forms of poorly controlled expansion which occur through the private sector, such as the spread of golf courses, which consume enormous amounts of land and water.
- Solid waste disposal – usually in the form of landfall dumping on unprepared, uncontrolled and ecologically unsuitable sites. Despite the presence of hazardous wastes, such sites often become a focus for the squatter settlements outlined above, as poor people seek to scavenge a livelihood out of recycling.
- Liquid waste disposal – this is one of the most common and serious impacts on the peripheral areas and beyond, as untreated sewage and industrial effluence enter rivers, lakes or aquifers, making agriculture or fisheries in the peripheral areas subject to contamination.

The impact of such problems on the urban periphery is magnified by the fact that it is usually an area of major importance for the urban poor, offering them opportunities for cheap shelter, woodfuel, farming and other economic activities. The destruction of this ecosystem is thus viewed differently by the urban and peri-urban populations.

Further afield the scale and impact of the urban footprint are becoming more marked every year. In the past, the economic base, and therefore the size of the city, was usually constrained by the resources of its hinterland, but the intensification of capitalism, together with the communications and transport revolutions has irrevocably broken the symbolic links between city and region. Much of the primary resource production of these regions is now exported, irrespective of the 'needs' of the city. Thus many African capital cities are surrounded by enterprises producing 'exotic' vegetables or flowers for European supermarkets while malnutrition is rising within those same cities.

Meeting the basic needs of the city had often brought fundamental changes to the surrounding area. Energy needs in particular have wrought devastating effects as biomass resources are exploited to the hilt (Plate 4.6). Woodfuel is still the major source of energy used in most African cities, so that regions for miles around have been

Plate 4.6 *Zimbabwe: woodfuel collection for urban markets*

denuded of trees by commercial retailers. Wood is brought into Harare, for example, from up to 150 kilometres away. When coal is a major energy source (domestically and industrially), mines and their waste dumps cover huge areas, while the burning of the fuel itself contributes to an air pollution problem that extends far beyond the city, both directly and indirectly, in the form of acid rain.

There are many other regional footprints which cities can make; the demand for building materials, for example. In Bangkok, this has resulted in the brickworks surrounding the Thai capital using up huge amounts of valuable top soil every year to satisfy the demands of that rapidly industrialising city (Parnwell, 1994). But perhaps the most difficult area of interaction between cities and their hinterlands relates to the supply of fresh water to urban consumers. The overuse and pollution of proximate sources, together with rapidly expanding demand from domestic and industrial consumers, and the lack of recycling, mean that water is being brought in from increasingly wide areas. Dakar, for example, draws much of its water supply from Lac de Guiers, some 200 kilometres away from the city. Cities located in semi-arid areas face particular problems and frequently suffer water shortages. Bulawayo, in Southern Zimbabwe, has faced such problems for years and is desperately seeking funds to tap the waters of the Zambezi. If these were ever to become available, what would the impact of such a drastic scheme be on the fragile ecosystem of this part of Southern Africa?

Urban management and the brown agenda

Are there any common denominators between the urbanisation process and environmental problems that would enable urban management to adopt appropriate policies? In terms of city size, there seems to be no obvious link with environmental issues. As ever, it is more a question of having the resources to cope and this would suggest that the most persistent problems will be in small, less prosperous settlements in poor regions or poor countries. However, pollution from industry and transport correlates positively with city size; moreover, the larger the city, the more extensive is its ecological footprint on the surrounding area.

Urban growth rates too have variable influence on environmental problems since the fastest growing cities are sometimes those with rapidly growing economies and are capable of making infrastructure investments. Admittedly, there are cities that grow quickly without parallel economic expansion. Urban densities too cannot be said to have a clear link with environmental difficulties. Some of the most densely populated cities in the world, such as Amsterdam and Copenhagen, have coped well with the pressures generated on their environments. On the other hand, some of the poorest quality living environments in the Third World (i.e. squatter settlements) often have low densities. Once again, it is the availability of resources and the willingness of the urban authorities to invest these resources which matters. Economic prosperity is, therefore, one area where there is a more obvious correlation with the management of environmental problems. As cities and their residents become wealthier so does their ability to transfer their environmental problems away from their homes and neighbourhoods to other people and other districts. For example, as garbage collection and sewerage systems improve, so the problem of waste is transferred from the household to the city, which in turn diverts such wastes to the communal sinks (rivers, marine or land). Although some environmental problems, such as air pollution from industry and transport, may worsen as cities grow more prosperous, their theoretical ability to deal with these improves. Singapore, for example, has stringent transport controls on car usage. Of course, while the nature of environmental problems may change as cities and their economies expand, urban management does not have to move in parallel with this change and can introduce improvements at earlier stages. The extent to which this occurs is not only dependent on the resources available, important though this is, but also on the nature of local political and administrative circumstances (and institutional assistance).

In some ways it is relatively easy to come up with a model of a planning process to respond to urban environmental issues (Figure 4.5), but putting such a plan into action is another matter. There are several factors that prevent appropriate policies being operationalised. The first is often a sheer lack of information and understanding of the contributory factors to the environmental crises, whether these are

Figure 4.5 *The urban environmental planning process*

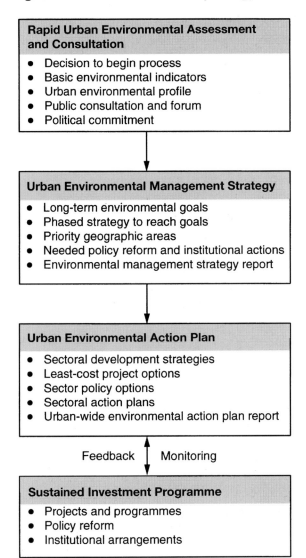

social, economic, political or technical. Second, and related to this general ignorance, is a lack of public awareness, other than through immediate, local experiences, of the overall extent of urban environmental problems. In many ways this suits the authorities, since protest is usually fragmented, small-scale and easy to contain. The third factor is the emphasis on economic growth at the expense of the urban environment. This produces weak regulatory systems, on the one hand, and poor support systems, for those who need this, on the other. In short, the urban management of environmental matters is poorly informed, poorly organised and poorly motivated. Above all, urban management clearly needs to recognise that environmental problems emanate as much from poverty as from unconstrained growth, so that appropriate responses need to address a wide range of social as well as regulatory matters.

As concerns have risen during the last decade, pressure has been growing at the global, institutional level for more positive policies. This has resulted in an unusual collaboration between the United Nations and the World Bank to establish the Urban Management Programme (UMP), a long-term technical co-operation programme focusing on five development areas: land, infrastructure, finance and administration, poverty and the environment. This programme, like many other contemporary development initiatives from global institutions, is driven and shaped by neo-liberal philosophies – a faith in the ability of economic growth and the market economy to underpin broader social and environmental changes. UMP likes to call these 'win–win' situations, in which economic and environmental goals are complementary. However, most of those firms involved in the productive process itself see little direct benefit to themselves in environmental expenditure and there is still an enormous role for the state to play. Unfortunately, the goals of many urban administrators and other elites in the Third World reflect the globalised green agenda of developed countries rather than the local needs of their own poorer citizens.

Some areas of management policy recommended by UMP do offer welcome changes of emphasis, particularly the need to raise public awareness, mobilise public support and involve local communities (in decision-making as well as in providing cheap labour). Other emphases are more contentious, particularly those that involve public–private partnerships. The incorporation of private sector companies is an important objective of many of these policies, since it reduces state involvement, a particular hobby-horse of market-oriented development strategies. At first glance, there would seem to be relatively little opportunity to involve the private sector in the brown agenda, but some success stories have emerged over the past ten years. This has particularly occurred in the field of solid waste collection, especially in Latin America and South-east Asia, where contracts to small private firms have improved services, reduced costs and effectively cleaned up large areas of many cities. However, not all areas of the city are affected because the service is based on another critical principle of the market economy – cost recovery. Low-income households that do not pay taxes or service charges do not usually experience improvements.

In the face of the retreat of the state from urban government and expenditure on infrastructural and welfare services, increasing emphasis has been placed on allowing and encouraging the poor to help themselves by the removal of constraints on the small-scale activities of the informal sector – 'building local capacity' as this is often termed. However, the poor already help themselves as best they can anyway and there are numerous examples of recycling of waste material (see Case Study L). For the most part, this is an uncoordinated response which does relatively little to improve the urban environment unless mobilised into community-wide action. But low-income groups are usually limited by their own resources, being able to contribute little other than their labour, unless the state itself is encouraged to input resource and capital, as is the case for aided self-help housing programmes. However, similar partnerships in the field of urban utilities, while not unknown, tend to be much more limited in scale and less uniform across developing nations.

In this context, there would seem to be considerable scope for non-governmental organisations (NGOs) to move into the gap between communities and government, injecting funds and organisational skills to co-ordinate labour and management into practical projects to improve the urban environment and help the poor. The ability and desire of NGOs to fulfil such roles have, however, been subject to considerable debate. Many NGOs comprise large, global enterprises and receive extensive subsidies from their own governments, making them as much if not more responsible and responsive to their superiors than to the communities with whom they work. But even successful NGO/community enterprises have been criticised for becoming part of a privatisation process for services or environmental change that ought, in practice, to be the responsibility of the state.

Certainly in the field of environmental issues, there appear to be more initiatives involving the poor than there are involving larger-scale contributors to the brown agenda. Regulatory policies are not in favour at present and many authorities prefer to develop and rely on economic instruments of control linked to some form of 'polluter pays' principle. However, with most governments keen to encourage economic growth above all else, and with considerable cross-interests existing between the commercial and political worlds, rigorous enactment and enforcement of such controls is still limited. At present, therefore, there appears to be a much greater burden of responsibility for urban environmental change being placed upon the poor.

Key issues

1. Urban environmental problems have a wide range of sources, from local to global.
2. Environmental pressures can result from irresponsible and uncontrolled development strategies. They can also result from poverty, itself caused by inequalities in the development process.
3. Within the household, access to clean water, sewerage provision, air pollution and overcrowding are particularly problematic, with direct consequences on health.
4. City-wide problems encompass air, water and ground pollution, and are often worsened by uncontrolled industrial and traffic growth.
5. The urban footprint in surrounding regions is deepening rapidly as wastes increase and resource demands grow.
6. Urban management of environmental problems is inadequate and often pushed onto the poor themselves.

Discussion questions

* Will the reduction of poverty help alleviate urban environmental problems?
* What is the role of urban management in meeting the brown agenda?
* Are health problems the result of ignorance or poor household management?
* What role can the private sector play in pollution controls?
* What are the special problems of the peri-urban areas with regard to environmental issues?

References and further reading

Asian Development Bank (1993) *Water Utilities Data Book*, ADB, Manila.

Bartone, C., Bernstein, J., Leitmann, S. and Eigèn, J. (1994) *Towards Environmental Strategies in Cities*, World Bank, Washington, D.C.

Bryant-Tokelau, J. (1994) 'Pacific urban environments', *The Courier (Fiji)*, 144.

Cook, I. (1993) 'Urban issues in the West Pacific Rim', paper presented to the ESRC Pacific Rim Research Group, John Moores University, Liverpool.

Drakakis-Smith, D. and Dixon, C. (1997) 'Sustainable urbanisation in Vietnam', *Geoforum*, 28(1): 21–38.

Elliott, J.A. (1999) *An Introduction to Sustainable Development*, Routledge, London.

Parnwell, M. (1994) 'Rural industrialisation and sustainable development in Thailand', *Quarterly Environmental Journal*, 2(1): 18–27

Pernia, E. (1992) 'Southeast Asia', in R. Stren, R. White and J. Whitney (eds) *Sustainable Cities*, Westview Press, Boulder, Colo.: 233–258

Rees, W. (1992) 'Ecological footprints and appropriate carrying capacity', *Environment and Urbanisation*, 4(2): 121–130.

Tevera, D. (1993) 'Waste recycling as a livelihood in the informal sector', in L. Zinyama, D.S. Tevera and S.D. Cumming (eds) *Harare*, University of Zimbabwe Press, Harare: 83–96.

UNCHS (1996) *An Urbanizing World*, Habitat, OUP, Oxford.

World Resources Institute (1996) *World Resources 1996–7*, WRI, Washington, D.C.

N.B.—*The Environment and Urbanization Journal* is an invaluable source of case study material.

⑤ Employment and incomes in the city

- ◉ Growth, development and urban employment
- ◉ Recent changes in the structure of urban labour markets
- ◉ Labour exploitation and response
- ◉ The informal sector
- ◉ Poverty in the city

Growth, development and urban employment

As Chapter 1 clearly indicated, the growth of urban populations is occurring at much faster rates than the growth of employment (Table 5.1). There are some problems with this correlation since the concepts of employment and work in the context of the Third World city are not the same as in the West (see Potter and Lloyd-Evans, 1998). However, it is true that the incorporation of urban populations into formal, waged work has been slower than their overall growth. Clearly this will vary from country to country but over most of the Third World urban jobs are difficult to obtain.

In many ways this flies in the face of conventional wisdom on the role of urbanisation in the development process since for decades urbanisation has been associated with economic growth, particularly

Table 5.1 *Comparative contribution of industry to employment and GDP*

Level of human development	GDP from industry (%)	Total employment in industry (%)
High	35	24
Medium	39	17
Low	24	13
Industrial	37	33

Note: 'Industry' comprises manufacturing in both formal and informal sectors, together with industrial primary activity.

Case Study M

Singapore: a model for development?

In 1965 when Singapore became an independent republic, the auguries for growth were unpromising. It had few resources, a crumbling built environment and had become the target for Chinese economic refugees from the new xenophobic states of the region. However, the Peoples Action Party, under Lee Kwan Yew, has completely transformed the economy into one of Asia's most impressive success stories.

Most of the early growth (Table M.1) was based on industrialisation, not only through manufacturing but also in heavier industries such as oil refining and shipbuilding and repairing. The great majority of this growth was funded by foreign investment or the state itself, while markets were predominantly in the USA and Europe. The Singapore government was one of the first in the region to recognise the limitations of low-technology industries and throughout the 1980s sought to transform the economy, through a 'second industrial revolution', to the 'Switzerland of Asia' with hi-tech, high-value industries, an extensive range of producer services and a regional financial centre. After a slow start this upgrade has eventually occurred. Singapore's low-technology industrial interests have been transferred to the 'growth triangle' (Figure M.1) in the nearby Riau islands of Indonesia and Johore state in Malaysia where Singaporean money and management have fused with cheap land and labour within extensive industrial estates.

These transformations have been based on opportunities provided by the development of global and regional economies, together with the perspicacity of the state, factors which are unlikely to be repeated elsewhere. Thus, Singapore cannot provide a model for development and certainly does not illustrate a free-market economy, as the World Bank would like us to believe. However, Singapore's economic success has also been heavily based on its one real resource – its people – and the consequence has been an attention to human resource management which has been very detailed and, some would claim, authoritarian and intensive.

Table M.1   Singapore: contribution to Gross Domestic Product

Industry group	1960	1970	1980	1990	1995
Primary	4.1	3.1	1.5	0.5	0.3
Manufacturing	11.9	20.4	29.1	29.4	26.5
Utilities	2.5	2.5	2.2	2.0	2.0
Construction	3.6	6.8	6.4	5.6	6.7
Trade	35.9	29.2	21.8	17.2	17.8
Transport/communication	14.2	11.0	14.0	13.3	15.2
Finance/business	11.3	14.0	19.6	28.0	27.6
Other services	17.4	12.7	9.1	9.9	10.2

Source: Department of Statistics (1996).

Note: Excludes bank service charges and import duties.

Figure M.1 SIJORI growth triangle

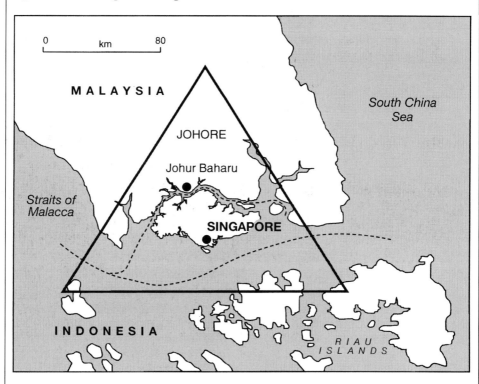

In the 1960s and 1970s this management largely took the form of strict population control policies in both fertility and migration, together with expanded and directed education and training programmes through which children were virtually allocated into 'hand or brain' streams at an early age. Tougher controls were also introduced on labour unions, and class solidarity was further fragmented by the introduction of factory- rather than occupational-based unionism. Company loyalty, Japanese-style, was also encouraged. Advancement was allegedly through ability in a self-styled meritocracy within a plural society in which all had equal opportunities.

However, disparities and exploitation have continued with Malays, in particular, still comprising the more disadvantaged ethnic group, with lower average incomes and larger families. Yet in numerical terms most of the poor are still Chinese and several sub-groups of vulnerable people have emerged in a society in which middle-class consumerism dominates. Migrant labour in particular, constitutes an exploited minority, while the rapidly expanding older populations too are finding life difficult in a state where welfare policies are limited and where great emphasis is placed on 'traditional Asian family values' in terms of taking responsibility for the aged. Such values have eroded over the years in spite of overt government encouragement. In addition to reducing family support for the elderly, such social changes have also resulted in an increase of single parents, mostly women, who not only receive virtually no support from the state but have been actively penalised – for example, in being excluded from eligibility for government housing.

There is no doubt that with an average per capita income which is higher than that for many European countries, Singapore has in a relatively short space of time created a strong economy which has brought substantial benefits to many in society. It is, moreover, a clean, green city which leads by example in its environmental policies. In social terms, however, there have been prices to pay for continued growth. Whether the economic crises of the late 1990s will challenge this situation remains to be seen.

through industrialisation, and the subsequent creation of employment. The state, therefore, has tried to promote such development in a variety of ways, from direct investment in production enterprises to indirect regulation and control over inputs, finance or planning. Such policies were deemed to be particularly important during the 1970s and 1980s when investment capital from multinational corporations (MNCs), banks, insurance companies and the like was seen to be flooding into the developing countries in search of increased profits.

It was thought that such investment flows were simply seeking cheap labour, but the demands of the global market for quality products has meant that other factors were important in conditioning the nature and direction of these investments. An educated and trainable workforce is particularly important in this respect, as are reliable and extensive infrastructural services, such as electricity and transport facilities. Few locations provided such opportunities in the 1970s, so that only a handful of countries experienced rapid growth, notably the four East Asian tigers (Hong Kong, Singapore, Taiwan and South Korea) together with Brazil and Mexico (Case Study M). Subsequently, this external investment-based growth has spread to other countries, such as Malaysia and Thailand. Ironically, much of the overseas investment for such growth is from the four tigers themselves which now boast their own MNCs (Figure 5.1).

Of course, investment was not in its entirety limited to so few countries. Both external and internal investment shifted in varying degrees to larger urban centres throughout the developing world. This tendency aggravated the already observable trend of focusing growth of all kinds into the largest cities in the urban hierarchy, creating the basis for the unequal regional development, primacy and mega-urbanisation that characterises so many developing countries today. Thus most manufacturing employment tends to be concentrated in these larger cities. For example, Greater Bangkok produces over three-quarters of manufacturing output in Thailand with very high proportions of new investment being channelled by the state through

Figure 5.1 *Waves of economic development in Pacific Asia*

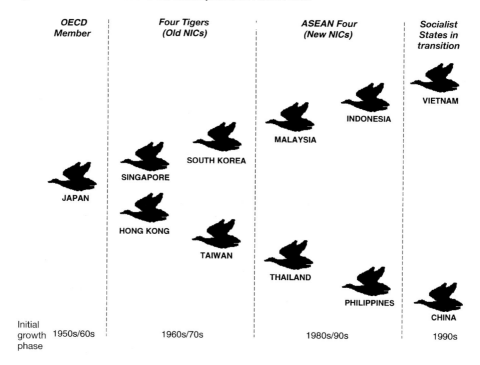

Whilst much of this manufacturing employment can be found in industrial estates, often in special export processing zones (with their financial incentives for MNCs) established by the state, most jobs are in small-scale enterprises, often barely 'formal' in their nature. In India, for example, almost 80 per cent of the industrial labour force is said to work in this sector (Chandra, 1992). Indeed, the small scale of manufacturing enterprises in Hong Kong and Taiwan has been claimed to be an important factor in their ability to adjust to market forces.

But not all of the employment found in large or small cities is in manufacturing. As Table 5.2 indicates, the service sector has always been the most important source of work. However, we need to be aware that this is a catch-all classification which encompasses a wide variety of jobs, from traditional marketing to employment in modern retail stores or international banks. Many of these jobs, particularly in capital cities or regional centres, are provided by the public sector. In many African cities, for example, it is claimed that 40 per cent of the urban workforce are employed in one official capacity or another. It is

Table 5.2 *Percentage contributions to GDP*

	Primary		Secondary		Tertiary	
	1970	1995	1970	1995	1970	1995
Low income economies	37	28	28	35	33	38
Sub-Saharan Africa	27	20	28	33	46	47
East Asia and Pacific	34	17	35	41	31	41
South Asia	44	30	21	26	34	44
Latin America and Caribbean	12		35		53	
	1960	1995	1960	1995	1960	1995
Low human development * economies	78	66	9	13	13	21
Medium human development economies	77	62	9	17	14	22
High human development economies	54	32	17	24	28	44

Sources: UNDP (1998), World Bank (1998).

* The Development Index as calculated by UNDP for developing countries only.

under circumstances such as these that structural adjustment programmes have had such a devastating impact on the urban labour force. Explicit to most structural adjustment programmes has been a drastic cut in government expenditure through the reduction of its own workforce. These had often been inflated simply to give educated men and women some form of income and occupation, usually through an established political and/or social network, in order to prevent them from fomenting unrest.

Recent changes in the structure of urban labour markets

The consequence of these economic and demographic trends, both short- and long-term, has been to create a fragmented labour market in the cities of the Third World. It is, therefore, a fundamental error to consider urban labour as simply one input into the equation of economic growth. As Connell and Lea (1994: 281) suggest 'new global economic arrangements . . . are shaped by the everyday practices of ordinary people'. Simple categories, such as formal and informal, while conceptually useful fail to capture the complexities of a situation in which households engage in complex and ever-changing work patterns simply to survive. Even hitherto 'safe' government employees are now feeling the pinch of recession or structural adjustment and many are involved in informal or illegal occupations, such as urban agriculture, to feed their families.

Two of the components of the urban labour market which have received much attention in recent years are women and, lately, children. Women have been incorporated into the urban labour market in a variety of ways according to local economic, social and cultural contexts. Common to most, however, is a channelling of women into the less-skilled and poorly paid jobs, the consequence of social stereotyping which itself is the consequence of a series of related factors (Table 5.3). Over large parts of the developing world, particularly in Africa, women are not extensively employed in factory work, whereas in Pacific Asia women now outnumber men in many of the urban manufacturing and service workforces. However, the benefit to women of this incorporation into the formal workforce has been debatable. Although wages become more reliable, they are often lower than those for men, even for the same tasks. Moreover, conditions of work for many women are extremely harsh, even to the extent of being locked into factories for the full working day. Many are young women, under 25 years of age, living in dormitories over the factory itself and limited in their social lives by restrictions placed on them by their employers. If living at home, moreover, many female factory workers are expected to continue to undertake domestic chores in addition to their waged work. Clearly there have been relatively few ideological gains to challenge existing patterns of gender inequality that have emerged from new work patterns. While it is clear that there is the potential for women to improve their status in society through

Table 5.3 *Factors affecting the disadvantageous position of women in the labour force*

1. Women's association with domestic and reproduction roles
2. Colonial establishment of the notion of the male breadwinner
3. Orthodox economic perception of women as low in education, skills and aptitude
4. Women envisaged as more passive, less organised and less likely to resist exploitation
5. Gender stereotyping of women as dextrous and able to undertake repetitive tasks better than men

Source: Potter and Lloyd-Evans (1998).

incorporation into the formal sector, this depends to a great extent on their age, status and family circumstances.

It would be a mistake to conceptualise the changing role of women in developing countries simply in terms of manufacturing. The level of growth, together with the creation of a consumer society and a middle class in the higher growth developing economies, has generated demand for additional labour in the construction, retail and service sectors. Much of this demand has been satisfied by movements from labour-surplus economies to labour-short economies. For example, there are estimated to be some two million Indonesians working in Malaysia, at least until the present economic crises made such migrant labour unpopular.

In this fluid labour context, one of the most rapidly increasing flows has been that of women into domestic service, particularly within South-east Asia where the prosperity of Hong Kong and Singapore has attracted workers from the Philippines, Indonesia and South Asia. Such flows are not new, nor confined to Pacific Asia. Indeed, much of our existing information relates to Hispanic women migrating to the United States. However, in Pacific Asia migration into domestic work has become an increasingly important strategy for countries unable to match the industrialisation rates of their neighbours. As with factory work, however, the conditions of employment and remuneration are still far from satisfactory.

Shirlena Huang and Brenda Yeoh (1996, 1998) have undertaken studies of domestic workers in Singapore that inform us of this situation. In the city-state the demand for immigrant domestic workers is a direct result of the increasing female labour force participation rate into the formal economy. This participation rate rose from some 20 per cent in 1957 to over half by the 1990s, and up to two-thirds for single women. The corollary is 80,000 foreign maids, three-quarters of whom are from the Philippines, who undertake the domestic tasks of the female Singaporean labour force. As the Singapore government is reluctant to become reliant on foreign labour both employers and employees face many controls to ensure that the workers are transient rather than permanent. Permits, in particular, are dependent on marriage and pregnancy restrictions which lead to employers exercising substantial control over their maids' social lives. In addition, the state provides little in the way of welfare legislation so that wages, conditions of work and personal treatment are the purview of the employer and maids can be dismissed at any time. Racism and class snobbery are not unknown factors in shaping such circumstances but only occasionally do examples of excessive cruelty or exploitation reach the press. Most maids dislike their position but remain for the money they can earn and remit to their families back home. Clearly this situation is one in which wealthy families exploit economically disadvantaged women in similar fashion to that occurring in many factories, a solution to a labour shortage that generates its own injustices.

Children constitute another group within the urban labour market with specific and special considerations that need to be taken into account. They constitute a varied component of the labour force, working in factories, as unpaid family labour, as scavengers, beggars and even as prostitutes in some Asian tourist destinations; one-third of the domestic servants in Indonesia are aged 15 or under. In this context, Morales-Gomez (1992: 5) notes that 'economic growth and social modernization in many Third World countries still have not created the changes necessary to recognize children as an important part of the development process'. In one sense, however, this can be regarded as a naive, if sympathetic, interpretation of the situation since the value of children's labour is only too well recognised and widely exploited. For example, the child labour force in Thailand is about 1.5 million (approximately the same size as the adult female labour force) of whom over one-third work in urban manufacturing industry in an 11- to 15-hour day and receive in return about half of the adult minimum wage. Silvers (1995), along with many others, reports even worse conditions in the carpet workshops of Pakistan and Northern India, with children bonded to carpet masters at 4 years of age and suffering severe physical handicaps as a result of their working conditions. Despite the concerns of international agencies and recommendations for change, there is little sympathy from employers. 'Our position is that the government must avoid humanitarian measures that harm our comparative advantage' (quoted in Silvers, 1995: 38).

It is true that children are often an important income earner in their family and are often not conceptualised as children, in the western sense, by poor families, but as small adults with responsibilities and obligations towards the household (Case Study N). Nevertheless, in certain countries, there does seem to be much scope in the workplace for consideration of the quality of children's lives without threatening their contribution to the family survival strategy. As with other areas associated with the labour force, things often improve through worker association. Boyden and Holden (1991), for example, cite the case of Vigilantes Mirins in Brazil who employ some 1,000 children in delivery services to 300 local firms in Brasilia, Belo Horizonte and Goiania. The children wear uniforms and hold membership cards but only deliver to firms sympathetic to their position. They receive the minimum adult wage for this work and are thus able to contribute much more effectively to their household budget. Of course, not all attempts to improve the situation are so successful. Silvers (1995) has charted the life of one articulate worker, Iqbal Masih, who became a leading light in the Bonded Labour Liberation Front in Pakistan. Despite, or perhaps because of, international recognition for his work on behalf of his fellow carpet-knotters, he was subsequently found shot. He was only 14 years old.

Many other groups in society are also exploited through the fragmentation of the labour market, particularly within the increasingly complex ethnic mixes which are being assembled in Third World cities as a consequence of the diversification and expansion of the migration field (Dwyer and Drakakis-Smith, 1996). In some countries, this ethnic diversity is spontaneously generated, such as in Bangkok where almost two-thirds of the migrants are from the Lao-dominated north-east region. In Malaysia, in contrast, increasing ethnic diversification of the urban labour force has been partly a consequence of the government's pro-Malay economic development policy which has encouraged a substantial shift of Malays from the countryside into urban areas in search of work. New tensions have been generated by the search for work by the Indian, Chinese and Malay communities.

Malaysia has recently experienced even more ethnic intensification of its urbanisation process through the expansion of immigrant labour, an increasing tendency throughout the Third World as transport costs decrease and perceptions of employment opportunities in rapidly industrialising cities grow. As noted earlier, South-east Asia has a particularly complex pattern of labour movements linked to the emergence of a hierarchy of labour skills (see Figure G.1). Thus, Malaysia is losing many of its workers to Singapore as they acquire new skills and values; in turn, it is experiencing labour shortages in unskilled areas, such as factories, domestic work and construction. Many of these vacancies are being filled by migrants from Indonesia who, being Muslim, are more acceptable to the Malay government. However, there are now an estimated one million Indonesians working in Malaysia, most of whom are illegal. The result is a rising resentment on the part of Malaysians of all ethnic groups. Elsewhere in the region there are also signs of an emergent racism in attitudes towards immigrant workers. Such issues of ethno-development have

Case Study N

Child labour in Tamil Nadu, India

Karur District in Tamil Nadu is a rural area dependent on rain-fed agriculture in which many children still work in both agricultural and industrial occupations. Altogether boys and girls make up some 5 per cent of the workforce. Why are such children in the workforce in Karur? Several principal reasons have emerged from a recent report and are reproduced here, together with case studies of children affected by them.

1 Poverty (47%).

   Nagaraju is 8 years old and has been working as a gem-cutter for six months, earning 5 rupees per day. His family lives in one of the driest areas of the state where farm work is irregular. His mother is a coolie and his father a blacksmith. He has two older brothers (13 and 19 years old), who work as coolie and carpenter respectively, and two younger siblings. The family has no land, lives in a thatch and mud house, and has borrowed from the gem-cutting owner in order to buy food. Nagaraju was, in effect, their collateral.

2 Lack of a culture of education (30%).

   This occurs when, despite having sufficient income, families still send their children to work. There is no real difference between girls and boys in this context, nor is it related to geographical accessibility.

   Angama has three children: a girl, 19 years old with no education, working in a mosquito net factory; and two sons, 13 and 15 years old, both of whom had completed their schooling to third standard only and also worked in the mosquito net factory. Angama and her husband have land, work and savings but she admits that she did not value education; they themselves were uneducated. Recently television had begun to raise her awareness of the value of schooling.

3 Discrimination against girls (19%).

   As girls enter another household after marriage, investment in their education is considered to be a waste of money. Often male siblings remain in school while girls are sent to work.

   Kavita was withdrawn from school at 14 and works in a garment factory in Karur. Her mother is a nutrition worker and her father a hand-loom weaver, both earning regular incomes. The family has a tiled house and land but also has a substantial loan from the weavers' society, taken out to pay for their house. In this case, the house clearly has a higher priority than the girl's education.

Source: Adapted from Lisa J. Freaney, 'Child Labour: A Case Study in Rural Tamil Nadu, India', Unpublished NGO report.

until very recently received little in the way of either conceptual or empirical investigation and yet an enhanced understanding of the relationship between ethnicity and economic growth is vital to an appreciation of the threats which may emerge to sustainable urbanisation. The recent explosive disruptions to so many societies, from the Balkans to central Africa, illustrate this only too well.

Labour exploitation and response

Common to all of these disadvantaged social or cultural fragments of the labour force is a general exploitation of workers, not only in terms of conditions in the workplace but also in terms of the social returns to labour, whether in the form of wages or welfare support. As Schmidt (1997) has commented, the market-friendly approach so beloved of the World Bank and neo-liberals is not so friendly to labour. Occupational health hazards are widespread and yet attempts to investigate these are often blocked by a combination of company and state interference. Sometimes the most guilty firms are the multinational companies. For example, in Bangkok, South Korean factories are notorious for their cruel treatment of Thai workers, and the settlement of one dispute included the proviso that South Korean supervisors were to 'stop hitting the workers even if they had explained something many times' (Heibert, 1993).

Along with appalling working conditions, most workers receive low wages and are supported by negligible welfare systems (either private or state). Even in successful countries such as South Korea, wage increases fell far behind the rate of growth of productivity, causing great hardship as the cost of urban living increased faster than the returns to labour. Despite World Bank claims (World Bank, 1993), therefore, poverty and inequality have been persistent even in rapidly growing economies such as Thailand and Malaysia (Schmidt, 1997). Welfare supplementation of low wages for urban households has been very meagre and tends to favour already-privileged groups, such as the government bureaucracy, MNC employees and the like. Attempts by the more enlightened western multinationals to improve conditions either in the factories they operate or those with which they have contracts, perhaps through some form of ethical investment programme, are often met with considerable opposition not only from the firms themselves but also from national governments. The Prime Minister of Malaysia, for example, claims that calls for a world-wide minimum wage constitute a blatant example of the West trying to 'eliminate the competitiveness of Pacific Asian economies' and sees it as disguised protectionism. The welfare state, as it has emerged in the West, is also seen as a growth handicap and the fact that few developing states have established such a system is seen as a comparative advantage. We should not be surprised, therefore, that large proportions of the rapidly expanding workforces of the ASEAN states remain uncovered by welfare provision (Table 5.4).

Responses on the part of labour itself to this situation have varied across time and space but in general have been very muted. Most governments continue to oppose labour organisations and only in a few instances have unions managed to break through this opposition. For example, during the Olympic Games in South Korea, global attention was used by trade unions to extract improvements in both wages and working conditions through strikes or threats to strike. Elsewhere, however, the situation has been less promising, even in 'socialist' states where the organisation of labour has deeper roots. Thus, the unions in Zimbabwe, once an important component of the liberation struggle, have effectively been silenced in the wake of structural adjustment. In Vietnam few of the new private companies have workers' unions and, as a consequence, wages often fall below stipulated minimum levels. In such circumstances it is usually regional MNCs or local firms which are worse offenders than western companies.

However, even these drawbacks do not discourage people from migrating into the city in search of work in factories or offices. Regular, waged work, even in exploitative conditions, seems preferable to no work at all. There is, however, an alternative for increasing

Table 5.4 *Welfare provision in selected ASEAN states*

	Non-covered labour force as percentage of total	Components of non-covered percentage			
		Wage earners	Self-employed	Family workers	Unemployed
Indonesia	88.0	13.7	41.9	29.6	2.8
Malaysia	65.8	24.9	22.6	12.3	6.0
Philippines	63.4	6.6	34.0	14.5	8.3
Thailand	89.6	17.8	19.6	34.3	7.9

Source: Schmidt (1997).

numbers of urban residents – that is, to seek work in what is commonly known as the informal sector. It is to this growing dimension of the urban labour market that we now turn our attention.

The informal sector

During the 1990s the promotion of small-scale enterprises has been seen as an important development strategy in both developed and developing countries. In the latter, however, such strategies are envisaged not only as important in their own right but also as an important adjunct to structural adjustment programmes and the unemployment they have created. Essentially these activities, whether retail or productive, are seen as straddling the gap between what has been called the formal and informal sectors of the economy but in essence their characteristics are overwhelming those of the latter and emphasise its importance to the evolving urban-based economies of developing countries.

The conceptual origins of the informal sector lie in the old dualism of the colonial period where two distinct economies and cultures were recognised. However, in the 1970s, as the failures of modernisation became more apparent, so the traditional or informal sector was recognised as playing a contrasting but important role in the economies of developing countries. This role was not only that of providing coping mechanisms for the poor themselves, for housing, food and jobs, but also the positive role it played in sustaining the formal activities of the urban economy and society – for example, the putting-out operations for many factories or domestic service for the wealthy.

In the 1970s and 1980s research into the informal sector became very popular and despite the difficulties of defining its activities, since individual and household activities are often complex and overlapping, it was revealed to be very extensive in many cities, providing employment for over half of the labour force in a wide variety of activities from scavenging on rubbish tips to repairing cars or motor cycles (Table 5.5). For many development strategists these revelations proved to be a godsend, since with minimal investment support the poor could be left to get on with their lives and meet their own needs at small cost to the authorities. Indeed, the links between the formal and informal sectors, which were subsequently shown to exist, were almost entirely to the benefit of the former (Figure 5.2). However, this is not to suggest that those in the dependent role did not

Table 5.5 *A typology of informal sector operations in developing countries*

....

Small-scale production

Primary activities	*Petty commodity production (secondary)*
Market gardening	Food processing
Urban farming	Home-production of hot food
Construction	Garments
	Crafts
Tertiary activities	Jewellery and trinkets
Printing and network	Shoes and leather products
Office equipment	Household goods
Computing and software	Electrical and mechanical items
	Specialised production (e.g. festivals, alcohol)

Distributive trades and tertiary enterprises

Distributive trades	
Processed food trading (nuts, snacks)	Clothes, shoes and leather goods
Unprocessed produce (fruit and vegetables)	Jewellery and cosmetics
Commercial food trading (Chiclet, Coca-Cola)	Newspapers
Suitcase trading (imported items)	Household items
Hot food and drinks	Music and electrical items

Tertiary services, daily providers	*Specialised services*
Laundry	Tourist guides
Domestic	Car park attendants
Show cleaning and repair	Car, home rentals
Hardware repair	Residential lodgings
Motor vehicle servicing	Secretarial, clerical
Taxi-driving and transport	Legal and medical
Maintenance and gardening	Beauty services/hairdressing
Odd jobs (e.g. car cleaning)	Distribution/storage
Bottle and waste collecting	Begging
	Protection

....

Source: Potter and Lloyd-Evans (1998).

benefit. The complex interpersonal nature of such activities provides some advantage to most engaged within it, enabling them to survive within the urban economy into which most have migrated (see Case Study O).

Potter and Lloyd-Evans (1998: 177) claim that the informal sector performs a variety of roles within society, encompassing 'wealth and

Figure 5.2 *Links between the formal and informal sectors*

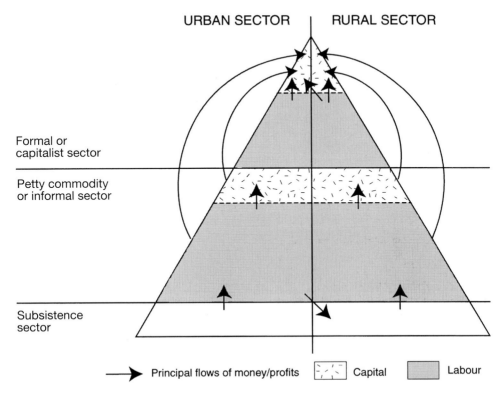

poverty, productivity and inefficiency, exploitation and liberation'
(Figure 5.3). Moreover, in the post structural adjustment era further
growth has occurred as migration to cities continues and employment
opportunities stagnate. Potter and Lloyd-Evans identify four
subsectors of informality:

1. Subsistence: where activities are geared towards self-consumption
   by the household. These increasingly include urban agriculture,
   even for the non-poor, with increasing job losses and food prices.
2. Small-scale producers and retailers: primarily an income-earning
   activity, which can be full-time or part-time. While production can
   be important, most activities revolve around retailing and most
   street traders fall into this sub-category.
3. Petty capitalists: these are largely small-scale production units
   which flout most regulations related to factory or labour
   conditions. As with the abuse of the environment, most
   exploitation of workers occurs within such activities since the
   producers are primarily concerned with profit.
4. Criminal: this is the undesirable and unacceptable face of the

Case Study O

Pondoks *and pedlars in Jakarta*

Petty trading, or informal-sector activities, in Jakarta is centred around *pondoks*. These are dwellings where the traders live and obtain their equipment and materials. However, they are not employees but self-employed residents of a sort of lodging house specifically for migrant traders in certain occupations who periodically return to their villages in rural Java. The *pondok* is run by a *tauke*, who is a cross between a landlady and an entrepreneur.

The *pondok* in this case study is run by a lady called Ibu Mus and its residents/workers all come from the same village and specialise in ice-cream making. The building is made of bamboo and various scavenged materials, and measures only 24 square metres but is home for some 15 people. Lea Jellinek, who lived nearby and studied this *pondok*, takes up the story:

> Life in the Mus household began at 4 a.m. when Ibu Mus and her husband Pak Manto received the day's delivery of ice . . . After the ice had been loaded into the cold storage they divided up the ice-cream ingredients which Ibu Mus had bought at the market the previous day . . . while Mus and Manto sat on the ground floor weighing out the ingredients (for each trader), the rest of the *pondok* awoke and began to descend the rickety ladder down to the ground . . . It was about 7 a.m. when they started rotating their buckets of ice-cream and they would sit there twisting and turning for the next three hours or so. Ibu Mus turned to her own work after she had finished weighing out the ice-cream ingredients. She carefully poured a variety of herbal medicines that she had prepared the previous evening into thirteen well-washed Johnnie Walker bottles . . . and by 6.30 a.m. set out to sell them. Ibu Sajum, Manto's sister-in-law, usually returned from market at the time Ibu Mus was setting out on her rounds and began to dice and cook the vast quantity of food she had brought back from market.
>
> It was still only 7 a.m. The ten ice-cream traders sat in a row mixing their ice-cream. Sajum was busy cooking. Manto occupied himself with all sorts of chores. Often Manto went to market and returned with an assortment of fresh fruit which he chopped into attractive little pieces and placed some ice fragments on top of them. When they were cold he would take them out on the streets to sell. But . . . on most days he devoted his talents to maintaining the ice-cream carts or carrying out alterations or repairs on the house.
>
> At about 10 a.m. the ice-cream was ready. Each trader tasted his product and made any final adjustments that he felt were necessary. Sometimes he asked his colleagues for their opinions and at other times he offered a sample to the group of children from the neighbourhood who invariably congregated when the ice-cream was nearing completion in the hope that their judgement would be called upon. When the product had been approved, the ice-cream bucket was carefully lifted into a push cart and surrounded by a fresh combination of salt and ice. Then, one by one, the traders strode into the kitchen, stripped and washed themselves [then] changed into clean singlets and shorts and sat down to the breakfast that Sajum had prepared.

But the breakfast had only been a small part of Sajum's cooking. She had produced a vast mound of fried savouries and titbits which she now neatly arranged on trays. Then she too changed into a traditional village sarong and kebaja and set off to sell her savouries. Soon everyone else followed suit. One by one the ice-cream traders manoeuvred their carts out of the narrow door of the *pondok*. Ibu Sajum had a regular clientele. Most of her customers worked in a big government office in Jakarta. They were the sweepers, messenger boys, guards and tea-makers. The office Sajum visited had its own staff cafeteria and traders were not welcome because they deprived the cafeteria of business. But Sajum sold cheaper food cooked in a traditional style and it was much in demand.

The ice-cream traders too had a regular route. The traders from Mus's *pondok* respected one another's territory and did not steal customers from each other. But, of course, they had to compete with the ice-cream sellers from other *pondoks*. Sajum's customers were buying their regular meal but the demand for ice-cream was rather more capricious and varied with the weather and the taste of the trader's ice-cream. The traders had invested a lot of money and labour into their ice-cream. As it was perishable and could not be refrozen, they could not afford to return to the *pondok* until their stocks had been completely sold. They usually kept to the narrow streets of the kampung, although the temptation of the major roads was always there. There were throngs of people out on the roads and many of them had rather more to spend than the people who crowded the back streets of the kampung. But there was a campaign against mobile traders and those who succumbed to the lure of the major roads risked losing not just their stalls but, even worse, their carts and all their equipment.

As Sajum and the ice-cream traders were beginning their rounds Ibu Mus's was drawing to an end. She too followed a constant route and in most places found regular customers for her herbal medicines which promised to combat a variety of ills ranging from infertility to unfaithfulness . . . If any of Mus's wealthier customers mentioned that they had some unwanted old clothing, Mus offered to relieve them of it and, if by chance others wanted to borrow money or buy some batik [cloth], Mus offered to help them with those needs as well.

By noon Mus returned home. She emptied out and thoroughly washed her thirteen Johnnie Walker bottles and stood them upside down to dry. Then she sat down to a meal which Sajam had left for her. When she had finished eating Mus set off for market. She returned with food for the evening meal which she prepared and cooked so that a meal would be ready whenever the ice-cream traders returned. When that was done she set off for the market again, this time to buy all the ingredients for the next day's ice-cream trade that she had been unable to carry on her earlier trip to the market. Mus found much to do about the house after the shopping was over. She carefully folded and stored away the old clothes she had collected. Later she would sell them in her village where they would fetch a fair price at festive times when the villagers felt they had to appear in a new, or at least different, set of clothing. Mus also collected any left over bread from the ice-cream traders who offered it to their customers as an alternative to ice-cream cones. Mus dried the bread in the sun and stored it away in glass jars. This too would fetch a reasonable price in the village when food was in short supply. By four in the afternoon Sajum returned from her food-selling and the two women set off together to collect instalments on the batik Mus had sold on credit as well as the money she loaned at an interest rate of 30 per cent per month!

From 6 p.m. onwards the ice-cream traders began to return. They looked exhausted as they pushed their way through the door of the *pondok* and parked their empty carts inside. They had started work at seven that morning and, if business was slow, it might be 9 p.m. or later before they began to make their way back home. Each trader unloaded and cleaned his cart and then silently consumed the meal Ibu Mas had prepared before climbing up into the attic and going off to sleep. They had neither the time nor the energy for socialising. The next day's ice would be delivered in a few hours and their work would begin all over again.

Source: This case study is extracted from Chapter 6 of *Food, Shelter and Transport in Southeast Asia and the Pacific* (1978), edited by P.J. Rimmer, D. Drakakis-Smith and T.G. McGee, Canberra, Australian National University.

informal sector, often linked to activities such as drug trading or prostitution. In many cases, however, there are major linkages to international crime or to the formal sector in terms of the overall structure, and the role that small-scale activities play.

Policies supporting informal sector activities, particularly the socially acceptable ones such as house-building, could be said to have reached a peak in the late 1970s and early 1980s within what has been called the Basic Needs Approach to development. The motives behind such policies were mixed, some were intended to ease the frustrations and the urban poor at relatively low cost, others were genuinely aimed at helping the poor help themselves. Whatever the motivation, many such policies were submerged in the rising tide of indebtedness and structural adjustment that has characterised development aid and development programmes since the 1980s and through the 1990s. Ultimately, however, the disastrous impact that structural adjustment had upon employment opportunities and basic needs subsidies has meant that the World Bank has had to develop socially acceptable adjustments and adjuncts to its policies. In part this was encouraged by the fact that those countries which were surviving recession were those where small-scale, flexible and often semi-legal enterprises predominated, such as Taiwan where no less that 92 per cent of all enterprises had fewer than 40 employees.

These policies should not be regarded as a dramatic reversal of attitudes – regulations controlling the informal sector have always been manipulated according to expediency. In developed countries the more formalised control mechanisms have given rise to flexible specialisation, based on multipurpose machinery and core labour force. In developing countries, however, it is more likely to be based on sweatshop labour and family enterprises so that deregulation can be

Figure 5.3 *Formal and informal sectors continuum*

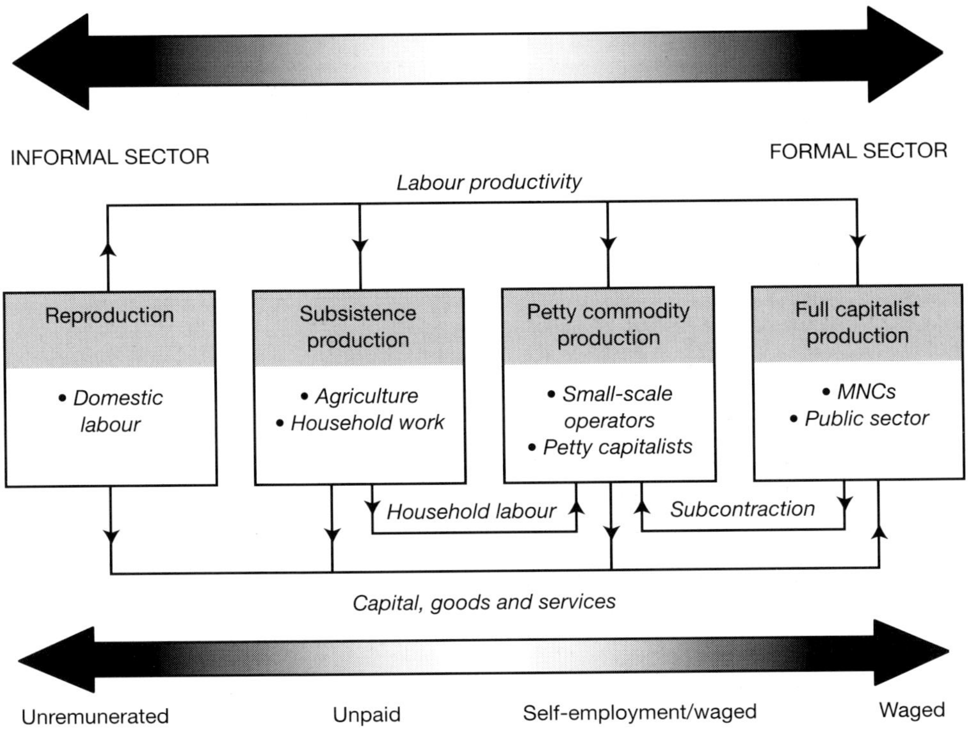

INFORMAL SECTOR   FORMAL SECTOR

Labour productivity

Reproduction	Subsistence production	Petty commodity production	Full capitalist production
• Domestic labour	• Agriculture • Household work	• Small-scale operators • Petty capitalists	• MNCs • Public sector

Household labour   *Subcontraction*

Capital, goods and services

Unremunerated   Unpaid   Self-employment/waged   Waged

Source: After Potter and Lloyd-Evans (1998).

seen to offer opportunities for further worker exploitation. Incorporating the 'dynamic informal sector' into the overall economy is clearly not as easy as it seems. For the great majority of its enterprises the informal sector does not equate with the flexible specialisation of the developed countries but rather a rather inflexible survival mechanism (Parnwell and Turner, 1998).

Thus, while there are many who do earn sufficient income from work in the informal sector to meet the basic needs of themselves and/or their families, there are many more who do not. In countries where state welfare provisions are meagre or non-existent the consequence is that large numbers of people in the cities of the Third World live in poverty.

Poverty in the city

The failures of successive development strategies to reach the poorest urban residents, coupled with the negative social impacts of structural adjustment, have seen a rise in the concern with growing urban poverty. As yet this growing interest has not generated much consistency in the debate on poverty, and there are particular problems with statistical data. The World Bank has claimed that poverty has been virtually eliminated from the more successful market economies, particularly those in Pacific Asia. However, other international agencies, such as UNDP, argue that while absolute poverty has diminished in these countries, relative poverty is still substantial, particularly as urban populations continue to grow faster than job opportunities. There is certainly no automatic and positive relationship between economic growth and poverty alleviation as the World Bank review of the East Asian 'miracle' would seem to imply. Indeed, as Table 5.6 reveals, unequal distribution of income is characteristic of all economies in Pacific Asia, irrespective of the level or pace of economic growth.

Despite these criticisms, World Bank data and optimism are frequently recycled in other authoritative institutional reports,

Table 5.6 *Income distribution in selected Asian countries*

| | Income share lowest 40 % | Ratio highest 20% to lowest 20% | Gini coefficient | Percentage of population living below poverty line | | |
				Total	Rural	Urban
Japan	22.3	4.3	—	—	—	—
Hong Kong	16.2	8.7	0.45	—	—	—
Korea	19.7	5.7	0.36	5	4	5
Singapore	15.0	9.6	0.42	10	—	—
Taiwan	20.0	8.0	—	—	—	—
Indonesia	20.8	4.9	0.31	39	16	—
Malaysia	12.9	11.7	0.48	16	23	8
Philippines	16.6	7.4	0.45	45	54	40
Thailand	15.0	8.3	0.47	30	29	7
China	17.4	6.5	0.31	9	13	—
Mayanmar	—	5.0	—	35	40	—
Laos	—	—	—	—	85	—
Papua New Guinea	—	—	—	73	75	—
Vietnam	—	—	—	54	60	27

Source: UNDP (1995), Asian Development Bank (1995).

becoming even more accepted in the process. For example, the Pacific Economic Cooperation Council, in an overview on poverty in the region, states that the Philippines has only 20 per cent in poverty, citing the World Bank, when almost all of the individual reports from that country put the proportion at double that amount. The review also virtually ignores urban poverty, suggesting that poverty increases with remoteness from the main centres and completely overlooking the fact that in almost all countries poverty is increasingly becoming concentrated in the main cities.

Clearly international comparisons of absolute poverty are almost impossible, even allowing for differential purchasing powers, since most countries set their own poverty lines, not always distinguishing between urban and rural poverty. As noted below, this glosses over some fundamental differences between the two. Table 5.7 lists the most recent figures for urban poverty. Clearly, some of these data are difficult to believe, those for China and Tunisia for example, and the many deficiencies of such estimates are discussed below. Notwithstanding such difficulties, however, the growing scale of urban poverty can be observed. In several countries, the raw proportion of those in poverty in the cities is already higher than the figure for the rural areas, traditionally the focus of concern. However, when the magnitude of urban populations is also taken into account, the total numbers in poverty in the urban areas are well in excess of rural totals for much of Asia and Latin America.

However, the condition of poverty is crudely expressed through poverty datum lines and usually reflects an external interpretation of the situation. Poverty is expressed through a fairly static measurement and fails to capture the fluidity and complexity of the lives of poor people. In all these restrictive income-based definitions, the poor are seen almost as passive victims and subjects of investigation rather than as human beings with something to contribute to both the investigation of their conditions and its alleviation. The poor often have quite different interpretations from outsiders about the particular problems they face. Rather than income levels or housing conditions, they place great importance on their vulnerability to sudden stress through insecurity. There is nothing startlingly radical in the contention that many of the causes of poverty are often long term, persistent and complex (Amis, 1995; Rakodi, 1995). What is relatively new and constitutes the driving force behind much of the new wave of concern is, first, the impetus and urgency given to the spread and deepening of poverty by the structural adjustment programmes imposed by the World Bank and IMF. The second impetus to this

Table 5.7 *Absolute poverty in urban and rural areas in selected countries*

	Percentage below poverty line		Urban % of total population	Ratio of numbers of rural poor to urban poor *
	Urban areas	Rural areas		
Africa				
Botswana	30.0	64.0	29	4.8
Côte d'Ivoire	30.0	26.0	41	1.3
Egypt	34.0	33.7	47	1.1
Morocco	28.0	32.0	49	1.2
Mozambique	32.0	70.0	28	4.5
Tunisia	40.0	5.7	55	0.7
Uganda	7.3	33.0	11	10.3
Asia				
Bangladesh	58.2	41.3	17	3.4
China	0.4	11.5	60	18.9
India	37.1	38.7	27	2.8
Indonesia	20.1	16.4	31	1.8
South Korea	4.6	4.4	73	0.3
Malaysia	8.3	22.4	44	3.2
Nepal	19.2	43.1	10	21.0
Pakistan	25.0	31.0	33	2.5
Philippines	40.0	54.1	43	1.8
Sri Lanka	27.6	45.7	22	6.1
Latin America				
Argentina	14.6	19.7	87	0.2
Brazil	37.7	65.9	76	0.6
Colombia	44.5	40.2	71	0.4
Costa Rica	11.6	32.7	48	3.2
Guatemala	61.4	85.4	40	2.1
Haiti	65.0	80.0	29	3.1
Honduras	73.0	80.2	45	1.3
Mexico	30.2	50.5	73	0.6
Panama	29.7	51.9	54	1.3
Peru	44.5	63.8	71	0.6
Uruguay	19.3	28.7	86	0.2
Venezuela	42.2	42.2	85	0.3

Sources: UNCHS (1996), World Bank (1994).

* In absolute totals: less than 1.0 indicates more urban poor than rural poor.

debate is the increased awareness of the spatial shift in poverty to more concentrated pockets in towns and cities.

As several observers have noted (see particularly Amis, 1995) poverty in urban areas is affected by a particular combination of factors which individually, while not unique to the poor, often combine to produce a special intensity and vulnerability. Perhaps most important in this context is the fact that the urban poor are much more immersed in the money economy. They must pay for virtually all their needs and consequently must have a cash income. Position in the labour market is, therefore, all-important, and the status and sector of employment is a strong determinant of income. Of course not all find work which brings regular hours or a reliable income and are forced into what is variously called the informal or petty commodity sector to manage as best they can with irregular or unreliable incomes. The corollary to low assets in this context is the higher expense involved in living in the city. Almost all basic needs cost more in the city and, ironically, are often more expensive per unit for the poor than for the non-poor. Food costs in particular pose problems for many of the poor, particularly in the context of increasingly expensive forms of retailing.

Structural adjustment has also made life particularly difficult for those living in cities. The removal of subsidies, particularly on basic needs such as housing or food, has hit those who were more reliant on them very hard and has made them more vulnerable to sudden stress and changes in their economic circumstances. Job loss has also been much more extensive in the city. Particularly important is the fact that hitherto stable jobs have been lost or experienced real wage cuts, increasing vulnerability enormously.

In coping with all of these difficult changes, the urban poor are also affected by a restricted range of strategies. For most, the option of return migration to their previous home is not a serious alternative, although this is happening in some areas, and rural assets, if still available, are often utilised (particularly with respect to food). Coping within the city is often difficult because of the more fragmented nature of the community and the household. Female-headed households are more frequent in the city and often lack the assets of partner-based households. In addition, community ties are weaker, making claim-based assets less prominent in limiting vulnerability. Physical assets too are, of course, more limited in the city. Poor urban households are very restricted in the extent to which they can grow food to supplement diet or income, or collect biomass fuel, although these vary with the physical environment of the city and its cultural/legal traditions. Thus urban agriculture is much more likely to be found in Zimbabwe than in neighbouring South Africa or Botswana.

The urban poor are also much more likely to suffer from environmental problems than people in rural areas. Those in employment frequently work in dangerous and/or polluted conditions, while the urban environment generally is usually subject to a much greater range of water and air pollutants (see Chapter 4). In addition, the urban poor come into much more frequent contact with the state and its agents, such as the police, experiencing to a much greater degree than rural dwellers various forms of corruption, harassment and other abuses of civil liberties (Wratten, 1995).

Despite these difficult conditions, the urban poor do try to cope as best they can with what is increasingly becoming a difficult life. One of the most consistent features revealed by studies of the poor over the last 20 years is that they are not passive. In an excellent summary of the responses by the poor to their worsening plight, Rakodi (1995) has outlined three (interrelated) strategies: those attempting to increase assets; those attempting to mitigate deteriorating consumption; and those designed to change household composition (Table 5.8).

Clearly the strategies currently adopted by households in, or vulnerable to, poverty must be the basis on which policy responses are formulated. Once again, this is not a new conclusion and many policies over the past two decades have attempted to do this; for example, aided self-help housing programmes. However, many of these policies have been too narrowly targeted and have sought to use the poor simply as a source of cheap labour. Current policy recommendations revolve around the view that not only are the poor dynamic but their poverty and vulnerability is multifaceted. However, most of the policy responses revolved simply around removal of the constraints on the informal sector and, as noted above, the capacity of the informal sector to continue to involute and create employment or to respond to initiatives has been shown to be limited. Usually this is related to the restricted capacity of labour to respond – for example, through physical incapacity, inadequate skills, time or transport constraints, poor access to credit and the like. These constraints are themselves, of course, the consequence of broader restrictions on access to assets, entitlements or resources.

This last point reminds us that almost all of the new policy recommendations can only become reality if appropriate and sympathetic institutional structures and operational procedures are

Table 5.8 *Urban household strategies for coping with worsening poverty*

....

Changing household composition

Migration

Increasing household size in order to maximise earning opportunities

Not increasing household size through fertility controls

Consumption controls

Reducing consumption

Buying cheaper items

Withdrawing children from school

Delaying medical treatment

Postponing maintenance or repairs to property or equipment

Limiting social contacts, including visits to rural areas

Increasing assets

More household members into workforce

Starting enterprises where possible

Increased subsistence activity such as growing food or gathering fuel

Increased sub-letting of rooms and/or shacks

....

Source: Rakodi (1995: 418).

put into place. Community-grounded poverty alleviation programmes become progressively more difficult with mega-city growth because it increases the social and often the physical distance between the poor and the policy-makers, while at the same time drawing away limited resources from smaller settlements. Yet mega-cities are precisely the direction towards which urbanisation in many developing countries is moving. Commanding legitimacy for more appropriate and sustainable poverty alleviation programmes in this context is not going to be easy.

How the rights and entitlements of the urban poor are expanded and protected constitutes a key issue within the debate on sustainable cities. If the urban poor continue to expand their numbers at such rapid rates, and if the capacity of the informal sector to absorb this increase is nearing saturation point, then more sympathetic and constructive policies must appear in the near future. Already the hardships resultant from structural adjustment have fuelled a desperation that has spilled over into urban violence and political instability, particularly in African cities. What avenues for voicing discontent and bringing about a real change of policy initiatives currently exist, and what will happen if these prove to be inadequate?

These fundamental political issues and related matters will be discussed in the final chapter of this volume.

Key issues

1. Rapid urban industrial growth has played a relatively limited role in Third World development.
2. Women have been incorporated into some urban labour markets but this has not necessarily transformed their status in society.
3. Children are widely exploited within the urban labour market.
4. Increased migration to cities has encouraged ethnic antagonisms in the labour market.
5. The informal sector is a large and extensive area of employment and has strong links to other dimensions of the urban economy.
6. Urban poverty is increasing and intensifying, often in spite of overall growth in the urban economy.
7. Absolute measures of poverty are less useful than indicators of vulnerability.

Discussion questions

* Why have more Third World countries not replicated the growth of the four Asian tigers?
* What benefits have accrued to women by their incorporation into the urban industrial labour force?
* Do children constitute a 'comparative economic advantage' and are objections to their employment designed to reduce competition to western manufacturers?
* Using a country of your choice illustrate how urban migration has resulted in deepening urban ethnic conflict.
* Has urban migration merely shifted poverty from the countryside into the city?

References and further reading

Amis, P. (1995) 'Making sense of urban poverty', *Environment and Urbanisation*, 7(1): 145–158.

Boyden, J. and Holden, P.C. (1991) *Children of the Cities*, Zed Books, London.

Chandra, R. (1992) *Industrialisation and Development in the Third World*, London, Routledge.

Connell, J. and Lea, J. (1994) 'Cities of parts, cities apart? Urban life in contemporary Malanesia', *Contemporary Pacific*, 6: 267–309.

Dwyer, D.J. and Drakakis-Smith, D. (1996) *Ethnicity and Development*, Wiley, London.

Heibert, M. (1993) 'Industrial disease', *Far Eastern Economic Review*, September 2: 16–17.

Huang, S. and Yeoh, B. (1996) 'Ties that bind: state policy and migrant female domestic helpers in Singapore', *Geoforum*, 27(4): 479–493.

Moralez-Gomez, D. (1992) 'Agents of change; children in development', *IDRC Reports*, 19(4): 4–6.

Parnwell, M. and Turner, S. (1998) 'Sustaining the unsustainable? City and society in Indonesia', *Third World Planning Review*, 20(2): 147–164.

Potter, R.B. and Lloyd-Evans, S. (1998) *The City in the Developing World*, Longman, Harlow.

Rakodi, C. (1995) 'Poverty lines or household strategies? A review of conceptual issues in the study of household poverty', *Habitat International*, 19(4): 407–426.

Schmidt, J.D. (1997) 'The challenge from Southeast Asia: social forces between equity and growth', in C.J. Dixon and D. Drakakis-Smith (eds) *Uneven Development in Southeast Asia*, Ashgate, Aldershot: 21–44.

Silvers, J. (1995) 'Death of a slave', *Sunday Times Magazine*, October 10: 36–41.

UNCHS (1996) *An Urbanizing World*, Habitat, OUP, Oxford.

UNDP (1998) *Human Development Report*, OUP, Oxford.

World Bank (1993) *East Asia Miracle*, Policy Research Report, World Bank, Washington, DC.

World Bank (1994) *World Development Report*, OUP, Oxford.

World Bank (1998) *World Development Report*, OUP, Oxford.

Wratten, E. (1995) 'Conceptualizing urban poverty', *Environment and Urbanisation*, 7(1): 11–37.

Yeoh, B. and Huang, S. (1998) 'Negotiating public space: strategies and styles of domestic workers in Singapore', *Urban Studies*, 35(3): 583–602.

6 Basic needs and human rights

- Introduction: defining basic needs
- Food security
- Shelter
- Human rights, democracy and urbanisation

Introduction: defining basic needs

As the previous chapter has clearly shown, there is unequal access to income within the cities of the Third World. As urban life is organised around the cash economy, this means that millions of citizens are going to have great difficulty in satisfying the basic needs of themselves and their families without help from the state. Unfortunately, over the past decade the imposition of structural adjustment programmes on so many developing countries has meant that the state has been forced to retreat from the limited welfare programmes that are in place, even in the larger cities where most state assistance is traditionally concentrated. In the face of continued urban population increase this means that the basic needs of most urban populations are falling behind what is desirable. Increasingly, the international institutions are looking towards the market to meet these needs, but how can the private sector reconcile meeting the demands of the poor while still making profits? And what role can the poor themselves play in meeting their own needs; a role increasingly emphasised by the World Bank to the problems its own structural adjustment programmes have created?

But what are basic needs? For the most part they tend to be used to refer to a wide-ranging set of issues with no common format. Many would include the infrastructural and service aspects of what are now usually separately addressed as the 'brown agenda'; others would include income and poverty. Such issues are not unwelcome additions

Figure 6.1 *The integrative nature of basic needs*

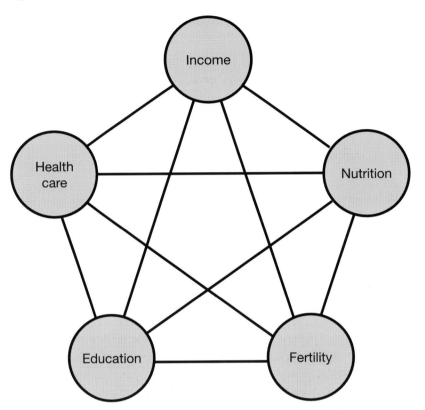

to the debate, serving as they do to emphasise the integrative nature of basic needs (Figure 6.1). In practice, however, there is a core of basic needs which most overviews address – namely housing, health care, education and food. Only recently has the last of these risen to the prominence it deserves within the basic needs debate, despite the fact that it dominates the expenditure patterns of the poor.

However, as with all aspects of sustainability, basic needs must be seen as an integrated whole. Tackling basic needs on a piecemeal basis, as has characterised most state responses in the past, is a self-defeating policy. The provision of an improved primary health care system, for example, is futile if those being treated return home to disease-ridden squatter shacks or are unable to meet the nutritional needs on which improved health depends. In theory, the state, the private sector and the individual household itself all have roles to play in meeting basic needs, but in practice the poor cope with their own needs as best they can on their own, with variable but diminishing assistance from the

Case Study P

Education in the city

In the last quarter of the twentieth century there has been considerable improvement in the provision of education within developing countries (Table P.1). Almost all actors in the development process favour investment in education as an important step in the development process – from the poorest peasant to global financial institutions. Moreover, in contrast to the past, when far too much money was invested in prestigious universities and colleges, the emphasis has shifted to primary education with almost 80 per cent of total enrolments now in this category. Many of the more successful Third World countries, such as the Asian Tigers, recognised this link between education and development many years ago. An educated and trainable workforce is of more value to an industrialist than a cheap labour force.

Table P.1 *Education provision in developing countries*

	SSA	Arab states	South Asia	East Asia	SE Asia and Pacific	Latin America and Caribbean	Least dev. countries	All dev. countries	OECD	World
Pupil–teacher ratio										
Primary	39	24	47	22	27	26	45	33	18	30
Secondary	25	18	31	15	19	15	26	22	14	20
Tertiary natural and applied science enrolment	30	26	26	44	23	28	23	30	21	25
Public expenditure on education as % of GNP										
1980	5.1	4.1	4.3	2.9	2.8	3.7	3.2	3.8	5.4	5.0
1992	5.7	6.4	3.8	2.8	3.4	4.2	3.0	3.9	5.4	5.1
Public expenditure on Primary and secondary education	79	74	69	73	76	64	78	71	72	72
Higher education	15	25	15	16	16	23	14	18	22	21
Literacy rate %										
Female	45	40	35	72	83	84	36	60	99	70
Male	65	66	62	89	91	87	58	78	99	83

Source: UNDP (1997).

But education does not always lead to economic success. There are countries which have impressive funding records but whose economies developed too slowly to absorb all the graduates, many of whom were accommodated in a burgeoning bureaucracy rather than being left unemployed, dissatisfied and a potential political threat. In Malaysia, for example, the federal bureaucracy swelled from 170,000 to 750,000 between 1970 and 1990, largely to provide jobs for the growing urban Malay workforce towards which the education system had been biased.

Education does not always move development along the path to equity. Most educational systems are western in structure and can often contain strong elements of anti-local culture, for example in religious schools. Indigenous educational systems are often repressed in order to impose a westernised uniformity on the development process. This is as true of Australia, where the educational system systematically repressed Aboriginal culture, as it is of many Islamic states in which Koran-based education is imposed on non-Muslim communities.

Education and urbanisation are strongly linked (Gould, 1994). In many countries the best educational facilities are located in cities, particularly capital cities, so that it is often necessary, or at least desirable, to migrate to these centres of education in order to become qualified. This is clearly exemplified in Bangkok which contains only 12 per cent of Thailand's population but 25 per cent of its primary students, almost 60 per cent of its secondary students and 85 per cent of the country's tertiary students. Once in the capital for education, few migrants return to their birthplace.

Much as the city is characterised by unequal access to other basic needs, so it is for education. Even prior to the imposition of structural adjustment programmes, schooling was never really free, even if no fees were charged. The cost of books, materials, uniforms and transport meant that many children from low-income households rarely had a full education. After structural adjustment, almost all countries were forced to introduce fees and reduce expenditure on schools and teachers' salaries. In Peru, for example, the 1980 budget was cut to two-thirds by 1990. Even small increases in the cost of education have enormous impacts on the poor and Figure P.1 shows a clear downturn in school enrolments in Vietnam after increases of only US 35 cents per month.

Many would agree, therefore, that investment in education has benefited the middle classes more than the poor and that this opportunity gap is widening. But within this class inequality, there is often another bias within the education system – namely, that against women. This discrepancy varies enormously but is most marked in Muslim states and also within secondary education. Often within poor households it is girls who are taken out of school first while their male siblings are allowed to continue their education. It is well known that such attitudes impede development as not only can educated women contribute more to the household economy but also become more influential in family planning decisions, usually to the benefit of the household. It must be added, however, that in many developing countries increased education for women is not reflected in their employment structure, with many in Pacific Asia, for example, simply being absorbed into factory work irrespective of their schooling.

In spite of these problems in the current urban system, it is vital that investment in education is maintained even within the structural adjustment programmes. More important still is that access for all low-income groups should be maintained. Basic needs are not only inextricably intertwined (see Figure 6.1), but in most societies they are underpinned by education.

Figure P.1 *Vietnam: education downturn after the introduction of fees*

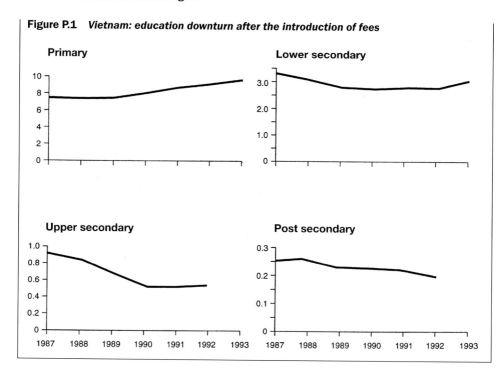

state, and minimal input from the private sector. This pattern is repeated no matter which basic need is being considered.

Meeting the physical and material needs of urban populations has rightly dominated most of the recent debate on improving the quality of urban life. No less important, however, is the question of human rights. Over and above the exceptional abuses which attract world attention, such as Tiananmen Square, there are daily abuses of human rights in the workplace, the street and the home, all of which play an important role in creating and sustaining inequality and vulnerability. We will review this situation later in this chapter, which will begin with overviews of shelter and food, together with shorter case studies of education and health (see Case Studies P and Q).

Food security

One of the main achievements of the last decade or so has been the increase in global food production, largely through industrialised farming in the West. Much of this cheap food has found its way into the cities of the Third World, in the process destroying much local agriculture and creating a dependence on imported staples, particularly wheat. Small-scale producers have been replaced by more

Case Study Q

Health and the city

Health issues in cities are largely a manifestation of poverty, as expressed in poor housing and sanitation, inadequate diets and lack of access to health facilities. In many ways the urban poor face a double set of health hazards – those associated with poverty, such as the communicable diseases like diarrhoea, together with the more urban problems of stress and poor working conditions. In effect they live at the interface between underdevelopment and industrialisation and suffer from both. It is misleading to think that urbanisation brings improved health because for most of the ever-increasing urban poor conditions are worse than in the rural areas. Investment in prestige health projects, such as hospitals, brings benefits only to the few who can afford to use them.

The vulnerability of the poor has also been increased by the structural adjustment programmes that have forced national and urban authorities to cut the subsidies on basic needs, including health services. Prior to the introduction of structural adjustment, there was considerably more variation in urban health care provision than there is now, when most have been reduced to a lowest common denominator. Figure Q.1 indicates how the imposition of small increases in health charges in Vietnam reduced the use of facilities on offer.

Figure Q.1 *Vietnam: health service usage after the introduction of fees*

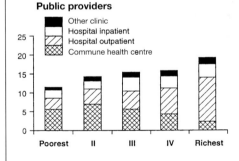

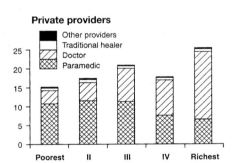

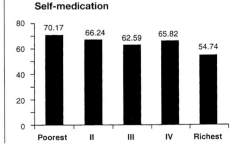

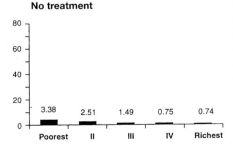

As migrants move into urban poverty they create environmental changes which themselves lead to the emergence of situations conducive to vector-borne diseases. Often such migrants bring with them diseases common to rural areas, such as schistosomiasis, and spread these around the peri-urban squatter areas in which they first settle. Indeed, environmental problems are a major cause of health difficulties in urban areas, including the tendency to overcrowding in both geographical and household terms. These can increase the risk of disease transmission, especially for infants (Figure Q.2), expose people to multiple infections and exacerbate the long-term impact of such exposures. HIV and AIDS, in particular, have paralleled the growth of urbanisation, particularly in Africa. However, the dangers of overcrowding are not always perceived in the same terms by the poor themselves who often have to make a trade-off between the increased health risks of crowding against the lowered housing costs and the reduction of equally pressing problems of debt or homelessness. Crowding in urban areas can, however, lead on to further health problems, for example in domestic accidents, particularly for women and children. Poor-quality living environments can also lead to acute respiratory difficulties because of poor insulation against cold and/or dampness, while lack of adequate ventilation worsens this risk by increasing indoor pollution from biomass heating and cooking facilities.

Figure Q.2 *The average disability-adjusted life-days lost per person per year from infectious and parasitic diseases*

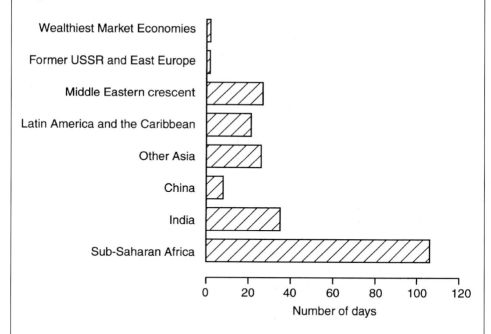

Perhaps the greatest health risk in most low-income urban areas comes from inadequate water provision and related sanitation facilities. These are closely interconnected since poor toilet and washing facilities usually lead to an accumulation of adverse environmental circumstances, such as the juxtaposition of open water storage and human waste, which foster the biological pathogens which cause diseases such as cholera, diarrhoea and typhus. Inadequate garbage collection facilities add to this problem, not only providing further breeding areas for disease but also by blocking drainage facilities during rains, spreading waste material even further afield (see Plate 4.5).

Work environments too pose many health problems in cities, exposing migrants to environmental circumstances which are new to them and with which they find difficulty in coping. Many factories, large and small, ignore basic safety regulations and workers are exposed to hazardous chemicals and wastes, as well as to excessive noise and vibration, much of which results in serious health problems such as skin diseases or respiratory difficulties. The psycho-social difficulties posed by long hours of work, few breaks, job insecurity and low pay also need to be taken into account in assessing health conditions in the workplace. Such psycho-social health problems are also found outside the workplace for many migrants, and would include exposure to excessive and uncontrolled traffic as well as to violence and crime.

As noted earlier, inadequate diets add to this myriad of health problems for the poor. Poor nutrition reduces the capacity to work and to cope with other pressures on health and urban life in general. Perhaps nothing illustrates better the inter-linkages between all basic needs as the increasing incidence of malnutrition amongst families in new low-cost housing schemes, as families struggle to meet rental demands by cutting back on food purchases.

Sources: Good discussions of health and urbanisation can be found in two volumes of *Environment and Urbanisation*: 5(2) 1993 and 8(2) 1996, both of which are largely devoted to this topic.

commercial units, often producing specialist crops for the urban market or even for export. Clearly urbanisation has played an important role in both the transformation of production and also in related access to food by consumers. Yet it is only in recent years that much interest has been shown in the relationship between urbanisation and food systems, a relationship which has important implications for our understanding of one of the most important basic needs in the city – namely, food security (Case Study R).

Conceptualising the urban food system is an important first step in more effective management of this essential element of urban sustainability (Figure 6.2). Food not only has crucial links to health, poverty and vulnerability, but also has important ties with employment, the urban environment and shelter. The great majority of urban dwellers buy most of their food, and this has given rise to an extensive but little considered range of employment opportunities. Cooked and uncooked food is retailed in Third World cities in a range of outlets, from casual street sellers through formal markets to shops, supermarkets and various fast-food sellers. Indeed, in the great majority of cities, food and beverage retailing dominate employment in the service sector, offering an extremely elastic activity that expands or contracts at the less formal end of the scale in relation to other earning activities. Food processing too, from simple cleaning and washing of fresh produce to cooking and canning, is often an extensive source of employment.

Case Study R

Food security in Harare, Zimbabwe

Zimbabwe inherited a productive but distorted economy when it became independent in 1980. During the first decade of independence economic growth rates were very respectable by African standards, but severe drought and the downturn in prices for Zimbabwe's primary exports combined to push the country's national debt to almost one-third of its GNP by the late 1980s. Almost inevitably this resulted in the imposition of a structural adjustment programme in Zimbabwe in 1991, known locally as ESAP. As elsewhere, ESAP opened up the economy to market forces, encouraging exports, reducing the bureaucracy and cutting back severely on state welfare subsidies.

The impact has been nothing short of disastrous for the urban poor, particularly as Zimbabwe's record of social investment had been good during the 1980s. Not only did unemployment rise but wages declined in real terms and the cost of basic needs grew substantially. Food security, as a result, has been threatened. Even in the early 1990s, despite rising agricultural production at the national level, some 80,000 to 100,000 people were being admitted into Harare's hospitals each year suffering from malnutrition.

Plate R.1 *Harare: urban agriculture*

As a result of ESAP the price of basic food rose rapidly. Immediately following its adoption in 1991 the price of maize meal, bread and milk rose by 14 per cent, 32 per cent and 13 per cent respectively. Between 1990 and 1993 the price index for bread and cereals had risen more than threefold, sparking off disturbances in several low-income suburbs. The escalating food insecurity also encouraged a much more widespread incidence of illegal agriculture cultivation on the many open areas of the city (Plate R.1). Initially, the reaction of the authorities was to invoke old colonial legislation and destroy these crops, but the widespread revulsion to this soon saw the cessation of such actions and urban agriculture is now more widespread than ever before.

However, the economic situation has continued to worsen in Zimbabwe and both inflation and the ever-tightening controls over public spending have continued to impact adversely on food security. By the mid-1990s almost 80 per cent of families in Harare had made changes to their diet involving fewer meals and/or decreased consumption of basic staples; consequential effects on health included an increase in the incidence of stunting amongst 1 to 4 year olds from 12 to 20 per cent. Throughout the 1990s consumer prices for food continued to rise relentlessly and by 1998 the consumer price index for maize meal and bread had risen to 700 from its 1990 base. In 1997/8, in particular, basic food prices rose dramatically as the value of the Zimbabwean dollar fell, with basic foodstuffs rising by up to 70 per cent in three months. Severe civil disturbances ensued which have begun to threaten overall political stability in one of Southern Africa's most viable economies.

Source: Drakakis-Smith (1995).

Both of these sets of activities relating to food are particularly important as sources of work and income for women and have recently been recognised as such in urban development programmes. Azami *et al.* (1996), for example, note that although food processing accounts for 17 per cent of all informal sector businesses in Bangladesh, there is further room for expansion in an activity that is suited to both decentralised and small-scale development, and that local and international NGOs have established some very effective training courses in this area.

In terms of urban management, this suggests that the removal of prejudices and constraints against such activities would be a sensible and positive way to support and sustain the quality of life of both vendors and consumers. Unfortunately, this is not what is happening in most cities where informal activities of all kinds, including those related to foods, are still subject to a range of proscriptive legislation and actions, most of which are legacies of the colonial period. In contrast, strong support has been given for what might be termed the westernisation of food retailing. This encompasses almost all aspects of food retailing from the origin and nature of the products themselves to the ways in which they are sold (Plate 6.1). Refrigeration and other forms of long-life processing, preparation and packaging have enabled supermarket chains and international food manufacturers to penetrate urban markets in the most unexpected places, so that such retailing dominates in cities from the South Pacific to South Africa, even precipitating 'store wars' between the large food

Figure 6.2 *The urban food system*

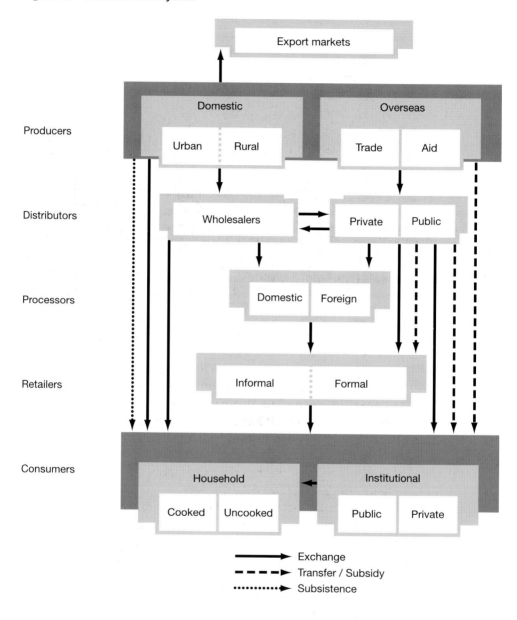

processing MNCs in countries with huge potential markets, such as Indonesia or Vietnam.

This process, and the parallel penetration by fast-food outlets (Plate 6.2), has been inextricably intertwined with the cultural transformation of urban populations in the Third World, particularly the trend towards western diets and even western foods. Of these the

Plate 6.1 *Harare: supermarkets in the suburbs*

most regrettable is perhaps the shift from local staples to bread, which often has to be produced with imported flour when suitable wheat cannot be grown locally – a legacy of both colonialism (Plate 6.3) and also of cheap exports of surplus US wheat during the 1960s and 1970s which effectively created economic and political dependency. The ultimate irony of much of this westernisation or 'industrialisation' of urban food systems is that it has, in some instances, created new health problems – replacing malnutrition with a range of morbidity patterns related to inappropriate or unhealthy diets. Perhaps the most unfortunate example of this has occurred in the US Pacific Dependency of the Marshall Islands where the cultural reorientation towards the United States has seen local products replaced almost entirely by imported convenience foodstuffs, so that although 'coconut, papaya and banana trees heavy with fruit, sway in the afternoon trade winds . . . kitchens are dominated by canned meats and rice, doughnuts and pancakes heavy with sugar. Local foods are valued less than imported goods' (*Pacific Islands Monthly*, August 1988: 44). The result has been a huge expansion in obesity and diet-related diseases such as diabetes.

Despite this relentless westernisation of urban food systems in the Third World, there has also been expansion of quite a different nature. This is the growth of urban agriculture which has occurred, on a widespread scale, but particularly in African cities where the real cost of retailed foods has risen substantially with the removal of subsidies,

Plate 6.2 *Ho Chi Minh City: American fast-food outlet*

the capitalisation of retail outlets and the fall in wage levels and employment. It is no coincidence that the 1990s has seen a plethora of studies on the persistence and growth of urban agriculture (see Egziabher *et al.*, 1994 and UNCHS, 1996). However, these empirical studies, while valuable as sources of information for urban planners, need to be placed into a firmer conceptual context if management decisions are to be properly informed. There is, for example, a reluctance to distinguish clearly between food grown in house gardens and production in illegal cultivation where the type of products, labour and material inputs are often quite different (Smith, 1998). There is also some confusion as to what the term 'urban agriculture' means. Many consider it to be a survival strategy adopted by the poor and, therefore, related primarily to self-consumption. However, there is also widespread middle-class involvement in urban agriculture, largely through its greater control over resource inputs. Furthermore, many low-income households grow food (legally and illegally) with the intention of selling it in order to raise additional income. However, through either saturated or inefficient marketing they are unable to do this and consequently end up consuming the produce. A final confusion relates to the distinction between fully commercial, small-scale urban agriculture and survival strategies. All of these different forms of urban agriculture need to be identified if more convincing policy recommendations are to be made.

Plate 6.3 *Vientiane: selling baguettes in a former French colony*

A further need for thought on urban agriculture relates to its role within sustainable urbanisation. In addition to the links with employment, vulnerability and health which have already been outlined, food systems are also integrated with other components of sustainability such as fertility and the environment. Illegal urban agriculture, for example, is often destroyed by the authorities on environmental grounds – for inducing soil erosion, silting or polluting hydrological systems, for example. However, as Bowyer-Bower (1995) has revealed, in one of the few ground-breaking exercises on this issue, much of the adverse environmental consequence of urban agriculture is manageable, given a supportive urban administration. But even less environmental damage would ensue if more households were encouraged to cultivate within their own gardens. Such a scenario, however, would involve forward and positive planning within the field of low-cost housing provision and, as yet, the role of formal garden space within the urban food and shelter systems has still to be thought through. Even here there are no obvious relationships as some families ignore the cultivation of garden crops in favour of flowers and lawns, on the grounds that the latter are more representative of the urban lifestyle to which they aspire.

Shelter

Housing poverty (Pugh, 1995) is often used to illustrate broader visions of poverty in the developing world. Frequently images are evoked of ramshackle, squalid shanty towns in order to evidence a variety of ills ranging from the evils of uncontrolled capitalism to the perils of uncontrolled fertility. Although in many cities the problems relating to finding adequate shelter are indeed worsening, such widespread generalisations often conceal the considerable efforts that squatters make to provide shelter for themselves and their families (Plate 6.4).

Data on the nature of housing poverty are readily available from innumerable empirical studies and in their local context these often present powerful and shocking scenarios. ESCAP, for example, estimates that, in spite of the region's economic successes, by the end of the century some 60 per cent of the Asian Pacific's urban population will be living in slums or squatter settlements (Pinches, 1994). Individual countries reveal equally pessimistic figures. Well over one-third of Egypt's urban dwellers live in unauthorised settlements characterised by inadequate construction materials and limited access to basic services, such as water or sewerage systems. The worst situation occurs in Greater Cairo where some 5.5 million people live in such settlements. Similarly in India the majority of cities in the high-density regions of the Ganges plains and the east coast have more than one-third of their populations living in slum settlements.

However Alan Gilbert (1992) has argued that international comparisons of data on shelter are bedevilled not only by unreliability but also by their subjective nature. What is acceptable as adequate shelter to a poor household in São Paulo may be quite different from that of a similarly disadvantaged family in Singapore or Lagos. The differing requirements of various points in the life-cycle of the migrant and/or low-income household will add further complexities (Figure 6.3). Thus for some groups proximity to employment may make relatively expensive inner city units with limited services acceptable whereas other residents in wealthier cities may demand much higher material standards. Even renting and ownership cannot be assumed to carry similar priorities. In some regions, such as West Africa, renting is far more commonplace and acceptable than in others where ownership is preferred. Such distinctions are very important since home ownership and tenure, in particular, are assumed to be fundamental basics on which successful self-help policies must be based. In this discussion, therefore, it might be more

Plate 6.4 *Hong Kong: squatter housing*

fruitful to move away from statistical and quantitative overviews and into an examination of the wider issues that are characteristic of the current debate on housing.

Many governments in developing countries have persistently refused to see the provision of adequate shelter as a priority issue in the development process. Low-cost housing provision, in particular, is considered to be resource-absorbing rather than productive and loses out to investment in industry or industrial infrastructure. There are, however, two dimensions to this debate. On the one hand there has been an intense discussion on the place that shelter provision ought to play within the development process as a whole (i.e. policies *towards* housing). For many years now this debate has been conducted at the global rather than the national level, with international agencies dominating strategy discussions with state governments. In parallel with, and often increasingly separate from, these events there has been another debate about practical policies *for* housing provision – the programmes and projects which are put into operation in the real urban world. For the most part, such activity occurs at a series of levels from the national state downwards to the citizens themselves. Over the years the nature of this debate has evolved around how the various agencies involved in shelter provision can best be combined within the delivery system to improve accessibility by the urban poor to better housing.

In the 1960s and 1970s the shelter debate was dominated to a great extent by discussions about housing provision – in particular, the merits and demerits of the various forms of aided self-help housing (Table 6.1) (see Potter and Lloyd-Evans, 1998, for an excellent overview). Many of the protagonists had had field experience of the poor and their housing needs, and the main disciplines involved were

Figure 6.3 *Housing and household life-cycles in the city*

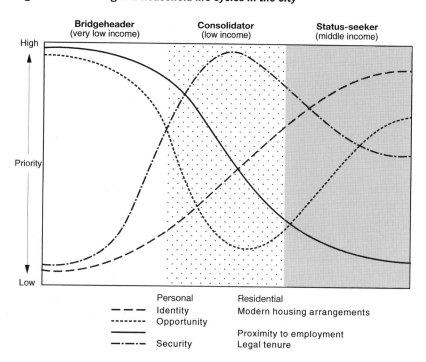

geography, sociology, architecture and urban planning. In many ways this debate has not been resolved and much of the literature on shelter still revolves around the same issues. It is partly for this reason that this area of discussion on shelter provision has been overshadowed in recent years by the macro-level policy debate on the place of housing provision within the broader context of the development process per se – a debate which has become much more dominated by economists and political scientists.

This shift began in the 1980s, partly in association with the imposition of structural adjustment programmes which saw a substantial retreat of the state even from the limited extent of welfare provision in which they were engaged. Housing provision suffered enormously and a major intensification of the urban housing crisis ensued in the face of continued urban population growth. The response of the poor was an expansion of informal housing – not only in squatter settlements but also, and more particularly, in renting (Case Study S). The reaction of the development agencies was to express a real concern for growing urban poverty, basic needs deficiencies of all kinds and potential instability. The result was a cluster of policies set within the 'new political economy' approach that ostensibly favours partnership and

Table 6.1 *Characteristics of informal and formal shelter*

	Informal		
Characteristics	*Early*	*Consolidating*	*Formal*
Residents' income	Low	Diverse	Middling
Builders	Users	Users/paid labour	Contractor
Mode of production	Artisanal	Mixed	Industrial
Value	Use	Commodifying	Exchange
Tenure	Illegal occupation	Mixed (inc. rent)	Mixed but illegal
Infrastructure	Little	Improvised/illegal	Provided
Legal status	Illegal	Mixed	Legal

Source: Adapted from Potter and Lloyd-Evans (1998).

integration between all the actors involved in housing provision in order to enhance the capacity of low-income households to improve their accommodation. However, as yet this is still a largely theoretical debate with only a few highly publicised successes.

Policies for urban housing provision

Aided self-help housing schemes emerged during the 1970s and 1980s as a major response to the perceived housing crises of the urban poor. Based on the sensible but belated realisation that the poor have a variety of positive objectives with regard to their own housing and on determination to achieve these, aided self-help housing involves some sort of collaboration between what have often been called the public and popular modes of housing construction (Figure 6.4). In theory, all parties benefit from joint-venture schemes that range from simple *in situ* upgrading to substantial core-housing projects, with improved housing and capital resources being obtained through a limited investment commitment by the state (Potter and Lloyd-Evans, 1998).

However, even aided self-help housing involves the investment of what may be limited national resources and many governments remain sceptical about investing in what they still regard as social overheads. If the enthusiastic support of international agencies such as the World Bank had not been available it is doubtful whether aided self-help schemes would have become as widespread as they did between the mid-1970s and 1980s. Even so, these schemes came in for considerable

Case Study S

Shelter in the city: Sebastiano's tale

Sebastiano and Maria used to live in a small village about 30 miles from Arequipa with Maria's parents. Work was difficult to obtain as the prices for the sugar that was grown in the area had been falling for years and only those with strong personal ties to the overseers were recruited. Sebastiano initially moved to Arequipa on his own in order to try to get work in the new factories in the city. He was unhappy at having to do this but Maria was pregnant and would be better off being looked after by her mother rather than roughing it in the city.

Sebastiano first moved into an inner city *barrio* with one of his distant cousins, but he only stayed there until he found out how the employment situation operated in Arequipa. Each day he would go to the western edge of the long-distance bus terminus where foremen from the building contractors would recruit their casual labour. Soon Sebastiano moved to his own rented room in a small house nearby. This accommodation was simple, even primitive, compared to his rural home. Basically he had a bed, a cupboard and a recess for hanging his clothes. Washing was done at the tap in the yard. His landlord was almost as poor as Sebastiano. He could not work because of his weak chest, the result of ten years' employment in the local tannery. The landlord rented out two rooms of his already small tenement, and he and his wife lived in only one room themselves, but they had first use of the kitchen and toilet.

For Sebastiano this was a good arrangement, his landlord did not demand the rent every week and understood that Sebastiano did not receive regular income from his casual work. Indeed, on the days that Sebastiano did not get employment, he worked with his landlord at constructing simple furniture which they made and sold on demand. There were also many sellers of cooked food near to Sebastiano's home so that he could get a good, cheap meal very easily. All in all, Sebastiano was reasonably happy with his rented room in the city centre – he was near to his main source of work, to cheap services and he had an understanding landlord. The advertisements for the new low-cost housing schemes on the edge of the city held no appeal for him. He could not afford the regular rental demands, nor the transport costs to get into the city to find work.

However, Sebastiano did not want to live alone in the city. He missed his wife Maria and their young son Pedro. After a couple of years in the inner city he began to make enquiries about a plot in a new squatter *barrio* that was planned on some unused public land. His landlord knew someone who knew the local councillor and advised Sebastiano to go and talk to him. The councillor was impressed with Sebastiano's carpentry and building skills and recommended him to the informal committee that was organising the 'invasion'. Sebastiano was accepted and with about ninety other families occupied the small piece of floodplain down by the river on the Ascension Day holiday when they knew the police would be busy elsewhere. Each family managed to put up basic walls and a roof on their allotted plot and, with the support of their councillor, were permitted to stay and improve their shelter as and when they could. They named their settlement St Christopher as they were all travellers to the city. The municipal authorities knew better than to enforce the letter of the law. After all, these were determined, hard-working people who were housing themselves at no cost to the city government (Table S.1).

Table S.1 *Opinions on aided self-help housing*

For	*Against*
● Many low-income households have obtained tangible benefits	● Low-cost housing is largely aimed at easing political frustrations of disadvantaged groups
● ASH programmes redistribute wealth by subsidising the poor	● There is no significant redistribution of wealth if rents/costs are too high. The poor default and are evicted in favour of the better off
● Some upward social mobility occurs by the shift from squatting to permanent housing	● ASH formalises a small portion of the informal section through the land and real-estate markets

Sebastiano had valuable skills. He quickly built a substantial shelter and moved his family to the city. He established a reputation as a carpenter who made good, cheap furniture – much in demand in this upwardly mobile community. Eventually he was forced to rent another plot in the settlement as a workshop and retail outlet. Sebastiano paid no rent, although he did make a contribution to the councillor's 'election fund' every month, and his business was slowly prospering; but he was concerned that he, and others in his community, had no legal title to their land. Eventually, however, sheer numbers forced the municipal authority to recognise the de facto rights of the settlement and to extend electricity, water and sanitation services to the *barrio*. Access to regular water was a particularly important improvement as residents had hitherto been forced to purchase from private water trucks at ten times the cost of the piped supply, but all of the upgraded services had resulted in a substantial improvement in the community's health and well-being. Sebastiano and Maria were now established members of the urban community.

criticism, largely on the grounds that they failed to fulfil their expectations. Poor management, higher than expected costs (and particularly the difficulty of recovering these), together with the failure to institute parallel and essential structural reforms in land markets, material supplies, building regulations and taxation systems, all contributed to this disillusionment, particularly for the international agencies.

More politically based critiques have also been made of aided self-help programmes. The use of the labour of the poor, without their parallel involvement in the broader planning and decision-making processes, led to not unjustified criticism of exploitation of the cheap labour of low-income groups. Ironically, aided self-help flourishes where governments need the political support of the poor or where a more organised yet cheap housing programme is needed to keep

Figure 6.4 *Typology of low-cost housing supply*

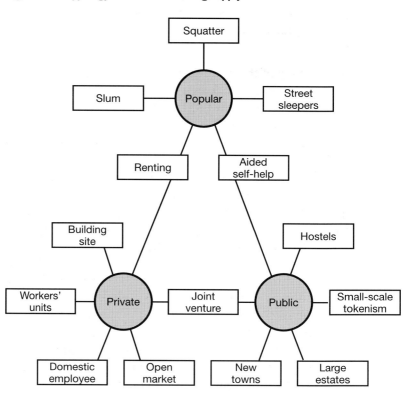

labour healthy but low cost. UNCHS (1996), in contrast, sees the main problem as increasing costs, resultant from inappropriate technologies or user prejudices, and considers more efficient management as the proper response. Irrespective of the debate on aided self-help housing, however, in reality the implementation of such schemes has steadily declined since the mid-1980s.

Although much of the current debate on housing provision, even in its diminished form, is still dominated by issues related to aided self-help it must not be assumed that other approaches are non-existent or are not being evaluated. Indeed, for a large proportion of the urban population of the Third World, the aided self-help debate has been largely irrelevant. Many socialist states exerted such strong controls over the urbanisation process as to preclude the growth of the massive unauthorised settlements and almost all low-cost housing was provided either by the state or within the existing rental market. However, as state socialism has retreated and controls over population movement have relaxed, so urban squatting has emerged as a prominent phenomenon, particularly in Africa. Elsewhere in the developing world many concerns still relate to more 'conventional' public housing schemes, such as large-scale clearance and redevelopment, that were being discarded decades ago in other countries.

While many of the overviews of low-cost housing systems which appeared during the 1990s have presented perceptive analyses of an evolving situation, it would be true to say that they represent a continuation of established and conventional evaluations. Few have sought as yet to place the examination of urban housing provision within the context of new concerns which have become increasingly prominent in development studies, particularly the notion of sustainability, perhaps because of the sheer scale and visibility of its 'problem' areas, such as squatting. Tipple (1996), however, examines ways of literally sustaining government-built housing through what he terms transformations, or extensions built for a variety of purposes, including those related to the home as a workplace. Islam (1996), on the other hand, while adopting a more conventional approach to an overview of housing problems in Bangladesh, inserts into this an assessment of the cultural and social issues involved, as well as the more usual economic and technological approaches. At present, however, the incorporation of housing into what might be termed broader discussions of urban sustainability usually relates only to infrastructural and environmental matters.

However, there are also more focused debates which revolve around particular themes within the housing–urbanisation–sustainability link. One such focus which has constituted an important part of the housing debate for some years, but particularly so in the 1990s, relates to sharing and renting. We still know relatively little about sharing, although it is a common practice, and few governments have formulated specific policies on renting other than simple rent control. It would be true to say that neither of these positions has changed substantially in the 1990s – despite the fact that the incidence of both sharing and renting has increased, as has homelessness (Table 6.2), due to a variety of constraints on both formal and informal housing construction in the face of continued urban population growth.

Such increases hold extremely important implications for housing policy, few of which are being recognised. Many of the attempts at community involvement in local improvement projects are unsuccessful because they are based on a false assumption of the desire for owner-occupancy and a related vested interest in community development. This failure to incorporate tenancy issues

Table 6.2 *Contrasts between poor housing and no housing*

Indicators	Squatter	Street homeless
Type of settlement	Spontaneous/informal	Spontaneous
Access to land	Invasion (legal/illegal)	Invasion/illegal
Permanence	Permanent/impermanent	Not permanent/itinerant
Organisation	Organised	Semi/not organised
Physical planning	Quasi/informal planning	Not planned
Location	Urban periphery	City centres (more often)
Growth	Increase/expands in density over a limited area	Expands without control
Type of building materials	Wood, iron sheets, sometimes mud walls are built	Scavenged cardboard boxes/ blankets
Life span of housing	Under five years	Weekly or monthly
Security of tenure	Partial or temporary	None
Safety	Minimum safety	Unsafe
Employment	Full-time employment in most cases	Unemployed/very erratic

Source: O. Olufemi (1998) 'Street homelessness in Johannesburg Inner City', *Environment and Urbanisation*, 10(2): 223–234.

into community development programmes is not for lack of information. Indeed, our growing knowledge of renting indicates that it is quite a diverse form of housing in its own right. Gilbert (1993) suggests that in Latin America most landlords operate on a small scale, have similar socio-economic characteristics to their tenants and are not exploitative. Recent research in Africa, however, indicates a quite different situation, not only with respect to the range of types of landlordism but also to the extraordinary degree of exploitation which occurs (Grant, 1996). Effectively what is occurring is a privatisation of housing supply, the benefits of which tend to filter upwards through a hierarchy of housing elites.

A second and perhaps more concentrated and rewarding focus for housing studies over recent years has been that related to gender issues. Often female-headed households, although in even greater need than most, are excluded from public housing schemes (Varley, 1996). This can occur indirectly, for example through level of earnings, or directly as a matter of policy. In Singapore, for example, female-headed single-parent households have been specifically denied access to an otherwise extensive and generous public housing programme, seriously disadvantaging such families (Davidson, 1996).

Entry into informal settlements too can be contained by inadequate income, access to credit or limited kinship networks.

The nature of the relationship between gender and housing is important in several respects – not the least of which is its impact on reproductive and productive labour. As the home is regarded by most men largely as 'feminised space', poor housing impinges on women's work in a wide range of ways, from the need to organise limited space for multiple uses, to the efforts required to obtain necessary basic services such as water. Women also contribute directly to productive labour through the home by renting rooms or extending domestic skills, such as taking in laundry or child-minding. However, even where women engage more frequently in formal employment, there are also specific needs which go unrecognised. It is often assumed, for example, that in the mega-cities of Pacific Asia or Latin America there is a pool of reserve female labour which has relatively easy access to manufacturing opportunities through the development of cheap transport, the implication being that housing needs are not an issue. However, as Yap and Rahman (1995) have revealed in their study of the Bangkok Metropolitan Region (BMR), this is not the case as most female factory labour is migrant labour and their housing needs are inadequately met by both the public and private sectors. The result is that most women are forced to live in rented rooms in informal housing subdivisions along the main roads leading out of the BMR, forcing their employers to run large fleets of buses that add to the already congested traffic of that environmentally challenged city.

Policies towards housing: strategic changes at the macro-level

In the wake of the ongoing urbanisation of poverty in developing countries, attention from the major development agencies has swung towards the city. However, this switch in attention and the impact that it has had on housing policy must also be set against the marked shift to neo-liberal development strategies in which the market was elevated to a more prominent role. Much of what little money has gone into the development of shelter programmes has, therefore, been made available by donors within strategies that have been laid down by the World Bank and the United Nations. Certainly the advent of neo-liberal policies, largely through structural adjustment programmes, has had a fundamental impact on housing provision as noted earlier. In Kenya, for example, Pugh (1995: 46) asserts that 'consequences for public sector investment in land, infrastructure and housing were immediate and adverse', and with few exceptions the situation was the same for most developing countries that were forced to adopt structural adjustment programmes.

The very crises that such policies triggered off, in the context of continued and accelerating urban growth, have been instrumental in the restructuring of macro housing policies that have characterised the 1990s. The World Bank, UNDP and UNCHS have begun to collaborate in their development programmes, with an increasing emphasis being placed on the integration of housing policies into the wider urban economy and management programmes. Pugh (1995) sees this as part of the new political economy (NPE) strategy which places much greater prominence on the role of the state. In this context the state can be seen to operate at two different levels, with urban policy being formulated at the national level in the context of both the national economy and international assistance. Operationalising this national policy, however, involves an enhanced role for the local state, in the form of the municipal authority, which is to co-ordinate the management of new housing policies and co-ordinate the partnership between the various local agencies (NGOs, individuals, or the private sector) which are to be involved.

Within this new partnership the emphasis is to be on 'enablement', facilitating the access of individual households to land, credit infrastructure and the like through the removal of restrictive legislation. This new global shelter strategy is, of course, to be at the expense of direct construction by the state, whether national or local. As Hesselberg (1995: 158) observes, 'the basic philosophy is that the poor shall be assisted to help themselves' and 'the initiative for shelter provision shall spring out of a more free and efficient market'. However, as many critics have noted, the most needy and vulnerable groups are rarely able to organise themselves and sustain collective action. Improvements will therefore depend not only on a sympathetic and genuinely facilitating local state but also on intervention by CBOs and NGOs. Enablement and partnership must not be seen as an excuse for public authorities to abandon their responsibilities in the field of housing provision, as many have been wanting to do for many years in order to favour economic investment. Certainly there is little in the way of extensive evidence that the new global settlement strategy has led, or might lead, to a transformation of housing policies in the real world.

In these circumstances, it seems, new ideas on shelter provision are unlikely to influence substantially the provision of housing for the urban poor for some time to come. It is perhaps understandable, therefore, that so many discussions seek to adhere to and build upon the admittedly limited, but tangible, successes of aided self-help schemes. In this context Ward and Macoloo's (1992) recommendations for maintaining the diversity of supply systems, easing supply bottlenecks, encouraging cross-subsidies, creating a greater role and accountability for the local state and decentralising discussion-making, still appear to be the basis on which positive change must be based.

Human rights, democracy and urbanisation

All of the individual components of sustainable urbanisation, from employment and clean water to adequate shelter, should constitute basic rights. These are not only integral to one another but are also inextricably linked to civil and political liberties, such as the right to vote, and to socio-economic entitlements, such as the right to education. Clearly such a body of rights is closely associated with urbanisation, so that effective management of the city must be underpinned by the search for equity as well as efficiency.

There is, however, no global agreement on human rights. Many Asian governments, in particular, challenge the universality of western concepts which are based on freedoms for the individual. Instead they argue for a set of communal values based on the family and mutual obligations. Some go further in their criticism of western human rights campaigns, arguing that they are partially designed to blunt the competitive edge of the Asian economics through minimum wage demands and improved working conditions.

In recent years some of the conditions relating to the extension of financial support through structural adjustment have been related to the adoption of 'good governance' (i.e a shift to 'western style' democracy). However, even in the West there is little agreement as to what democracy means (Table 6.3). It is not surprising, therefore, that there is little correspondence in the developing world between democracy and development. In practice, democracy is not only plural but also fluid, changing its characteristics over time in response to a variety of internal or external stimuli.

The World Bank would, of course, argue that what its strategies encourage is participatory democratisation by empowering individuals through freedom of choice in a more liberal market

Table 6.3 Types of democracy

	Radical democracy	Guided democracy	Liberal democracy	Socialist democracy	Consociational democracy
Objectives	Enabling undifferentiated individuals to exercise their rights and protect their interests	Achievement of the 'general will'	Representation and protection of diverse interests	Equality, social justice	Consensus between diverse groups
Role of the state	Executor of the will of majority	Executor of the general will	Referee	Redistributor of resources and guide action	Referee
Political process	Provision of arena for pursuit of individual interests	Unchecked pursuit of objectives proclaimed by the ruling elite	Checks and balances to prevent tyranny of the majority, or its representatives, or powerful minorities	All citizens given an equal voice by reducing inequality of wealth and resources	Recognition of the diversity of interests and identities by bringing leaders of all major groups into the governmental process
Citizen participation	Active participation is encouraged; electoral contestation	Mobilisation by ruling elite; no elections to key institutions, or only non-contested elections	Permitted but not actively encouraged; electoral contestation	Popular participation to offset elite power; may involve mobilisation or coercion	Participation within constituent groups, and by group leaders in the allocation of resources; electoral contestation
Actual and potential problems	Tyranny of the majority	Tyranny of the elite	Elite domination on account of unequal distribution of resources	Extent of coercion required to achieve objectives	Reinforcement of social division

Source: Pinkney (1993).

economy. However, the reality is that most governments do not address the problems of 'the poor' unless that group is recognised as having some collective political clout. Munslow and Ekoko (1995), in particular, take issue with what they term 'the mirage of power to the people'. They are sceptical about effective decentralisation because the financial, human and technical resources are usually not readily available, and argue that reinforcement of civil society does not guarantee improved democracy. 'At its best, participatory democracy is really about a transfer of power and resources, not to people directly, but to NGOs and other representatives at grassroots level' (ibid.: 175).

Within this realist position, we must now ask what relationship such trends have with urbanisation. Much of the recent discussion on political decentralisation seems to ignore the process of rapid urbanisation. Given the fact that rapid urban population growth and urban-oriented economic growth are dominant features of development, it might be expected that the urbanisation process would have some impact on the nature of democracy and human rights. However, the literature rarely addresses such matters directly; usually there are only allusions to the impact of urbanisation. From time to time, specific human rights are addressed because of their urban context, such as the call for the right to shelter in the 1990s. All too often the urbanisation process is integral to the discourse but is seldom addressed as such.

There is little doubt that there is some positive correlation between the level of urbanisation and support for human rights, although there is also perhaps a reverse correlation between rate of urban growth and human rights. However, these relationships are not strong and wide variations exist, largely because urbanisation is common to a range of political systems. Botswana and Kenya, for example, have the same relatively low levels of urbanisation and high rates of urban growth but vastly different human rights records; the same can be said of Costa Rica and Gabon at higher levels of urbanisation. Notwithstanding this weakish relationship, it is evident that the urbanisation process must focus and intensify the struggle for both democratisation and human rights, however these are defined. Earlier sections of this book have indicated clearly that most urban residents are denied access to adequate levels of income, to basic needs and also to proper channels of representation and individual human rights. In this context, of course, they may be no worse off than their rural counterparts. What is different in the city, whether large or small, is the concentration of deprivation, the opportunity to associate in order to protest, and the proximity of the perceived targets for such complaints. It is no coincidence that the sharpest protests in favour of greater democracy, representation or redistribution have occurred in capital cities such as Beijing, Caracas or Jakarta, whatever the political system.

There are, of course, many ways in which specific urban interests can protest at their disadvantaged position and seek to make their lives more sustainable. These may range from the individual and formal, within properly franchised and secret elections, to more group-oriented and informal activities, such as squatter land invasions. While most of those involved in both covert and overt, formal and informal, actions are usually the underprivileged, leadership of such action may well originate elsewhere through opportunist political agitators, self-interested patrons or even other class groups. The role of the emerging middle class in demands by urban interests for improved human rights or democratic procedures is currently being debated. However, the point is surely that the urbanisation process has facilitated the emergence of such classes and is, therefore, playing an important role in both the discourse and reality of political change. In similar fashion the urban industrial context provides the potential for the emergence of trade unions which may, or may not, seek to advance the economic, social and political rights to their members. Not all trade unions seek to pursue such goals, particularly if they have effectively been established as an extension of state control, as in Singapore for example, but most do seek to extend the rights of their members against both employer and governmental opposition, even when the latter may be ostensibly left-wing.

The urbanisation process has also intensified protest over human rights and democratisation by virtue of the facilitating agencies and organisations that tend to be based in the city. This would include international and domestic NGOs of all kinds, together with less formal community-based movements. Of course, not all of these theoretical means of empowerment result in widening the base of political participation but most would seek to improve access to improved social and economic rights, if not political rights. Certainly, the fact that cities, particularly capital cities, contain the targets of such pressure groups – namely, government offices, international and national agency headquarters or banks reinforces the role of the city and urbanisation in democratisation and increasing access to human rights.

Key issues

1. All basic needs, however these are defined, are strongly inter-linked so that improvement programmes should reinforce each other.
2. Food retailing throughout the Third World has become westernised in content and style.
3. State housing programmes have generally diminished over the years.
4. In many cities renting is as important as squatting to low-income households as a housing strategy.
5. Most basic needs policies are strongly influenced by external forces.
6. Western-style democracy is not always welcomed by, or suitable for, all developing countries.

Discussion questions

* Why are basic needs so poorly and erratically met within the Third World?
* How do the urban poor attempt to meet their own food needs?
* How can the special needs of women be incorporated into low-cost housing programmes?
* Are squatter settlements a problem or a solution?
* How does debt affect basic needs provision?
* Why has the introduction of democratic government not suited all developing countries?

References and further reading

Azami, S., Brough, S. and Battcock, M. (1996) 'Food processing and urbanization in Bangladesh', *Appropriate Technology*, 23(1): 9–11.

Bowyer-Bower, T. (1995) 'The environmental implications of urban agriculture in Harare', Paper presented at a Workshop on Urban Agriculture in Harare, University of Zimbabwe.

Davidson, G.M. (1996) 'The spaces of coping: women and poverty in Singapore', *Singapore Journal of Tropical Geography*, 17(2): 313–331.

Drakakis-Smith, D. (1995) 'Food systems and the poor in Harare under conditions of structural adjustment', *Geografisker Annaler*, 76(B): 3–20.

Egziabher, A. *et al.* (1994) *Cities Feeding People*, IDRC, Ottawa.

Gilbert, A. (1992) 'Third World cities: housing, infrastructure and servicing', *Urban Studies*, 29(3–4): 435–460.

Gilbert, A. *et al.* (1993) *In Search of a Home*, University College Press, London.

Gould, W.T.S. (1994) 'Literacy: school enrolments: higher education', *Atlas of World Development*, Wiley, London: 144–151.

Grant, M. (1996) 'Vulnerability and privilege: transitions in the supply of rental shelter in a mid-sized Zimbabwean city', *Geoforum*, 27(2): 241–260.

Hesselberg, J. (1995) 'Urban poverty and shelter: an introduction', *Norsk Geografisker Tidskrift*, 49: 151–160.

Islam, N. (1996) 'Sustainability issues in urban housing in a low-income country: Bangladesh', *Habitat International*, 20(3): 377–388.

Munslow, B. and Ekoko, F.E. (1995) 'Is democracy necessary for sustainable development?', *Democratisation*, 2: 158–178.

Pinches, M. (1994) 'Urbanization in Asia', in L. Jayasuriya and M. Lee (eds) *Social Dimensions of Development*, Paradigm, Sydney: 114–121.

Pinkney, R. (1993) *Democracy in the Third World*, Open University Press, Buckingham.

Potter, R.B. and Lloyd-Evans, S. (1998) *The City in the Developing World*, Longman, Harlow.

Pugh, C. (1995) 'The role of the World Bank in housing', in B. Aldrich and R. Sandhu (eds) *Housing the Urban Poor*, Zed Books, London: 34–92.

Smith, D. (1998) 'Urban food systems and the poor in developing countries', *Transactions of the Institute of British Geographers*, 23: 207–219.

Tipple, G. (1996) 'Housing extensions and sustainable development', *Habitat International*, 20(3): 367–376.

UNCHS (1996) *An Urbanizing World*, Habitat, OUP, Oxford.

UNDP (1997) *Human Development Report*, OUP, Oxford.

Varley, A. (1996) 'Women-headed households: some more equal than others', *World Development*, 24: 505–520.

Ward, P. and Macoloo, C. (1992) 'Articulation theory and self-help housing in practice in the 1990s', *International Journal of Urban and Regional Research*, 16: 60–80.

Yap, K.S. and Ramhan, A. (1995) 'Housing women factory workers in the northern corridor of the Bangkok Metropolitan Region', in T.G. McGee and I. Robinson (eds) *The Mega-Urban Regions of Southeast Asia*, UBC Press, Vancouver: 133–149.

7 Planning and management

- Introduction
- External influences
- National and local influences
- Conclusions

Introduction

Previous chapters have indicated clearly that urban growth and development constitute an extremely complex process and, in general, are accompanied by a whole range of problems which stretch across the economy, society and their environment. Devas and Rakodi (1993) argue that leaving the urban development process largely to market mechanisms is unrealistic and unsatisfactory since large numbers of low-income residents and many low-profit activities are only marginal to the market, while the unequal distribution of wealth will prevent many households and individuals from accessing those resources and services which may be available. They suggest, therefore, that 'the question is not whether the state should intervene, but rather to what extent it should intervene, and what form that intervention should take' (Devas and Rakodi, 1993:32). Table 7.1 summarises the scope of such intervention.

A role for government or the state implies that urban development can and should be managed or planned. To undertake such tasks urban managers have a range of instruments available to them which may be loosely classified as regulatory and financial mechanisms plus direct intervention. But several different actors may be involved in management – politicians, administrators, professional planners, consultants and donor agencies – all of whom may come into conflict over the methods and goals of urban management. When perceived in such a way, planning and management can become very elitist,

Table 7.1 *The scope for government intervention in the urban system*

Area	Activities
Protection of the public	Law and order; protection of human rights and property rights.
Regulation of the private sector	Administrative controls; tax and pricing systems; land and building regulations; pollution controls. Usually undertaken by the national state.
Provision of public services	Infrastructure provision; housing programmes; basic needs, such as education, health care, etc. Often municipal but national and private roles are common.
Redistribution of wealth	Taxation and subsidies; welfare programmes; wage regulations, tenure rights. A mixture of national and local processes.
Production	Means of production (land, labour and capital) may be controlled by the state (national or local) or the private sector.

Source: Adapted from Devas and Rakodi (1993).

Note: There is often conflict between the various branches of the state and the market over these activities and controls.

top–down processes, derived from the West and passed across to developing countries during the colonial and post-colonial periods. The whole concept of 'master plans', which still exist for hundreds of cities in the Third World, reeks of such elitism and raises serious questions about urban management in such countries. Who controls the planning and management processes, what are their goals and objectives and what are the means by which these are achieved?

In many ways this poses an even broader set of questions than may first seem to be the case. There is not only the question of whether citizens or planners shape the urban development process, but also whether the planners/managers are municipal, national or foreign in their roles and loyalties. Examination of the processes involved in urban management over the last two decades reveals many contradictory forces in operation.

External influences

At the global level the major international development agencies have for many years invested their funds in rural development during the very decades when major shifts of population into the cities have been occurring. Most urban investment schemes were fragmented, front-loaded, and western-influenced planning exercises that brought little in the way of long-lasting improvements to the cities involved. However, by the late 1980s and early 1990s it was becoming apparent to these international agencies that cities were not only the principal engines of economic growth in developing countries, but that they were also potential centres of political turmoil and unrest as urban management failed to come to grips with their enormous social and environmental problems. Under such conditions not only was urban sustainability under threat but so was the sustained growth to which the economic fortunes of the West were so closely linked. A quote from a special issue of *Time* magazine in 1993 summed up the mood of the moment: 'with urban areas producing about half of the world's income and governments nervous about restive urban populations, agencies such as the World Bank have begun to focus more on cities' (p. 34).

Early in the 1990s both the World Bank (1991) and the United Nations (1991) produced important policy statements on urbanisation and its problems. Perhaps more importantly they have come together in their Urban Management Programme with the specific goal of strengthening the capacity of urban managers to meet the challenges posed by rapid growth. In perhaps too many instances this has led to aid programmes targeting management and planning at the expense of practical projects to improve basic needs directly. However, there is no doubt that the capacity to manage has been improved and this is nowhere more apparent than in the Healthy Cities Project of the World Health Organisation (WHO). WHO considers that sustainable urban health is a complex process linked to all other aspects of sustainable urban development. Indeed, a model to appreciate this relationship (Figure 7.1) bears a close resemblance to that of sustainable urbanisation in Chapter 1. WHO aims to promote urban health by involving all of the actors and agencies within a city to agree jointly, despite differing perspectives, on what they might do to improve the living environment and health care. Initially begun in Europe in 1986, the Healthy Cities Project has been adopted in recent years by a variety of cities in developing countries.

In many ways this capacity-building role has extended external

Figure 7.1 *A framework for the consideration of urban health*

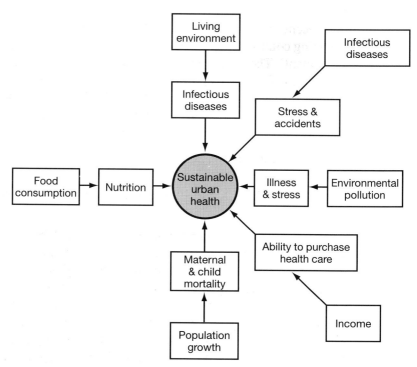

Source: Adapted from T. Harphram and E. Werrna (1996) 'Sustainable urban health in developing countries', *Habitat International*, 20(3): 421–430.

influence over the development process in Third World cities – not only are many economies controlled by the structural adjustment programmes mentioned so often in this text, but the management of big cities has also been influenced by the goals of the same international agencies. Ironically, this increased external control and capacity-building has been accompanied by the retreat of the state from investment in subsidies and welfare programmes of any kind. As we will see later, the response to this, encouraged by the same external agencies, has been to throw the responsibilities for welfare services upon the poor themselves (Gibbon *et al.*, 1992). The poor are being encouraged to use their own initiatives to create employment, housing and health care and many other basic needs from which the state is to retreat. *Time* magazine described this as 'the awakening of self-reliance in the poor'. Of course, such self-reliance has always existed but has been rediscovered on 'useful' occasions. It is in this light that we need to review the alleged shift towards local participation in urban management that has occurred in the 1990s.

National and local influences

Many recent discussions of national and urban management in developing countries prefer to use the term 'governance' rather than 'government'. The former means more than the latter, 'encompassing the activities of a range of groups – political, social and governmental – as well as their inter-relationships' (UNCHS, 1996:161). Three factors have allegedly helped local governance to emerge as a key arena within which urban management must be set:

- Increasing democratisation
- Increasing decentralisation
- Increasing importance of community involvement in decision-making

Decentralisation takes a variety of forms (Table 7.2) but in general involves the delegation of authority from higher to lower levels of state administration, from the national to the regional or local. This shift has been encouraged by the World Bank on the grounds that local service needs vary and are better understood at the local level and may be better integrated with one another. However, success in this localisation process depends heavily on the willingness of central governments to devolve both power and financial capacity to municipal authorities – something many are reluctant to do (Case Study T). Indeed, in many countries the management of large capital cities is shared, if not dominated, by the national governments which

Table 7.2 *Forms of decentralisation*

- *Deconcentration*, or the transfer of functions, but not power, from a central unit to a local administrative office. This is one of the 'weakest' forms of decentralisation and has become a common response by higher levels of government to deflect the blame for inadequate services provision from central to local authorities.

- *Delegation*, which involves, in most cases, the transfer of certain powers to parastatal agencies of the central state. While the parastatals have a certain autonomy in day-to-day management, they are usually controlled ultimately by government.

- *Devolution*, considered by some as 'real decentralisation', since power and functions are actually transferred to sub-national political entities, which, in turn, have real autonomy in many important respects.

- *Privatisation*, which involves the transfer of power and responsibility for certain state functions to private groups or companies.

Source: UNCHS (1996).

Case Study T

Two Latin American experiences of decentralisation

In Ecuador, although the number of municipalities has increased dramatically during the last decade, they have progressively lost functions and powers to the central government because much of the oil wealth has accrued to the central government. This is despite the fact that the military period ended in 1978 and a new constitution established the basis for democratic government. Municipal governments are responsible for the provision of water and sewerage, solid and liquid waste disposal, public lighting, control of food products, land use, markets, tourism promotion, and authorisation for the functioning of industrial, commercial and professional services. As in many countries in the region, the capacity to manage urban issues within municipalities is weak both in terms of human and financial resources, and it is characterised by ineffectual and obsolete administrative systems and weak social participation.

The transition to democratic rule in Chile began with the gradual strengthening of municipal governments, and increasing their level of resources. This was followed in 1981 by the transfer to municipalities of the responsibility for the administration of education and health. The next step was the modification of the constitution to allow for the direct election of local councillors and mayors. Municipalities are responsible for public transport and traffic, urban planning, refuse collection, parks, promoting communal development, local roads, sewers, public lighting, managing health-care centres, and primary and secondary education. They are not responsible for the provision and treatment of water nor for regional public corporations. They share responsibility, with the central government and other public institutions, for social assistance, environmental protection, public housing, drainage, and support to special social programmes.

Since the advent of the democratic period in 1988, Chile has embarked on a decentralisation programme which has the potential of significantly strengthening the ability of local governments to play a more decisive role in urban governance. This decentralisation has included both administrative and political aspects with the intention of enabling a more direct involvement of civic organisations in daily political life. Through various fiscal measures, real resources for municipalities increased by 36.1 per cent between 1985 and 1991. In terms of urban governance the potential for greater civic participation is certainly evident and likely to increase as the democratic administration of municipalities is further entrenched.

Source: UNCHS (1996: 167–168).

are located within them. Thus national sectoral policies and ministries determine the housing, educational and health care programmes in cities such as Kuala Lumpur and Seoul. The rise of large secondary cities is, therefore, creating new conflicts.

Decentralisation has allegedly been facilitated by the second of these

Case Study U

The commercial real estate market in Russia

The commercial real estate market in Russia is organised in a way that encourages corruption. Local governments own most commercial real estate; in 1995 the proportion was approximately 95 per cent. The struggle for real estate assets is part of a broader conflict between the central and local governments. The subnational governments are winning. Even when the central government attempts to require the property to be privatised, local governments simply refuse to obey, decreeing that federal law has no force in their community.

Ownership rights are vested in local councils, which are legislative bodies. As a practical matter, however, the head of an administration has a great deal of personal influence in the management of real estate assets. One reason for the strong local authority is that the laws and decrees enacted by central authorities are ambiguous, leaving the rules unclear. In some local governments the committee for the preservation of architectural and historical monuments also has considerable power, since almost all buildings in the city centre are on the register. Although its formal powers are unclear, in practice these committees often have veto power over property sales. Housing maintenance agencies must give their consent before a lease or sale occurs. Local bureaux of technical inventory maintain property records. Although they lack formal authority, their monopoly over technical information gives them leverage. Other regulatory agencies with influence over the use of property are the land reform committee, the fire department, the sanitation and health department and the district administration. The lack of standard procedures combined with the weakness of leases means that these agencies feel little pressure to co-operate or to refrain from imposing additional demands after a lease is signed.

Commercial real estate allocation does not follow commercial principles. Present occupants are favoured, and rental rates are far below market prices. The user's relationship to the head of the administration is often a factor determining rental rates. Officials use their control over urban real estate not only to enrich themselves but also to subsidise organisations such as social clubs – charities that obtain space cheaply and then use it for commercial activity.

Reform will be difficult. The problem is not just the incentives for graft, but also a tradition of heavy government involvement in real estate. The stream of personal benefits that accrue to city officials clearly plays a large role in perpetuating non-commercial management. City officials benefit personally from controlling real estate allocation, in terms of both money and influence. But an important and related dynamic is often missed. Russia has a strong tradition of extensive government involvement in economic activity on every level. This tradition, which predates the Soviet era, leads both officials and citizens to expect that the government will actively intervene in business operations and economic development. While many of the formal institutional mechanisms for this practice were removed in the initial stages of transition, several levers remain: taxation, regulation and control of real estate.

Moscow and St Petersburg have overcome the diffuse ownership structure in commercial real estate. These cities have a special enclave status that gives them greater legislative and administrative freedom. With access to outside technical assistance, they have restructured the management of their real estate assets. Unfortunately, this has simply permitted city managers to exploit their monopoly power more effectively. The city is both primary developer and regulator, leading to conflicts of interest and favouritism towards 'friendly' developers. Officials work for both private developers and the city, and many receive a good apartment or shares in a development company.

Source: Harding 1995, in UNDP (1997) *Corruption and Good Governance*, Discussion Paper 3, Management Development and Policy Division, Bureau for Policy and Programme Support, UNDP, New York.

factors, the increase of democratisation in the 1980s and 1990s. Democratisation in this context is, however, largely interpreted through western eyes as a system in which representatives are selected through fair, honest and regular elections in which candidates compete freely for votes and in which all the adult population is eligible to vote. As the previous chapter has noted, however, there are many different forms of democratisation (see Table 6.3) and the insistence on the adoption of western democracy as a condition for loans has not led to a smooth development process over many parts of Africa, for example, where tribal or regional rivalries have been woven into the democratic process to produce less rather than more spatial democracy.

In particular, the decentralisation of democracy to the regional or municipal level has not proved to be as closely related to the introduction of democracy at national levels as was hoped. Mayors are still too often appointed by national executives, while municipal authorities are rarely elected on the open, broadly franchised lines recommended for national governments. It is in this context that the 'partnerships' so favoured by recent urban development strategies need to be placed, since weak or unrepresentative municipal authorities can easily engage in the corruption or cronyism that continues to favour limited groups within society and business (Case Study U). It is to counteract such trends that recent years have seen a rise in interest in participatory politics – the incorporation of citizens voices into a bottom–up planning strategy.

Participatory approaches essentially developed out of research techniques which sought to give citizens a greater voice in the decision-making processes affecting their lives. As Table 7.3 indicates, such approaches can run the full gamut of effectiveness, from mere lip-service to situations where the participants initiate actions themselves. In theory, these approaches can increase the flow of

Table 7.3 *Different types of participation in urban planning*

....

Self-mobilisation	Project initiated by the population themselves who develop contacts with external institutions for resources and technical advice they need, but retain control over how resources are used. Within low-income settlements, this is less of a project and more part of a process by which people organise to get things done and negotiate with external agencies for support in doing so.
Interactive participation	Project initiated by external agency working with local population (and often in response to local people's demand). Participation seen as a citizen's right, not just as a means to achieve project goals. People participate in joint analysis, development of action plans and formation or strengthening of institutions for implementation and management. As such, they have considerable influence in determining how available resources are to be used.
Functional participation	Participation seen by external agency as a means to achieve its project goals, especially reduced costs (through people providing free labour and management). At its worst, external agency's goals accord little with the people's own goals. However, this limited form of participation can bring real benefits to 'beneficiaries' in many instances.
Participation for material incentives	People participate but only in implementation in response to material incentives (for example, contributing labour to a project in return for cash or food; building a house within a 'self-help' project as a condition for obtaining the land and services).
Consultation and information-giving	People's views sought through a consultation process whose aim is to elicit their needs and priorities; but this process is undertaken by external agents who define the information-gathering process and control the analysis through which the problem is defined and the solutions designed. No decision-making powers given to the population and no obligation on the part of the project designers to respond to their priorities.
Passive participation	People are told what is going to happen but without their views sought and with no power to change what will happen.
Manipulation and decoration	Pretence of participation – for example, with representatives of the 'people' who are not elected and have no power.

....

Source: UNCHS (1996).

information for planners and enable residents to prioritise their needs and to identify development goals, and the local resources to assist in their achievement. Often, however, the consultation process is restricted to 'key' figures in the community with a corresponding limitation in beneficiaries. In most cases, functional participation (see Table 7.3) is the dominant process, with citizens being used simply as cheap labour to cut the cost to the state.

Little wonder, in circumstances such as these, that participation has over the years become much more politicised and is organised through various citizens' groups or urban social movements. In many cases the incentive to organise is linked to particular issues, such as squatter demolitions or escalating crime, or to particular groups within society, such as women. Often mobilisation is short-lived and groups disband after achieving their objectives, or in the face of concerted opposition by the state, but they constitute an invaluable part of the planning process by attracting the attention of the authorities to crisis points within the fabric of the city.

Urban social groups have their origins in the 1970s when the number of squatters began to build up considerably in Third World cities. In the face of policies designed simply to eradicate such settlements, squatters formed resistance groups of various kinds, some of which were able to make the transition into participatory planning, giving rise to a variety of aided self-help programmes (see Chapter 6 and Case Study S). Frequently this transition from protest to participation occurs with the assistance of intermediary organisations that facilitate contact between community and government. One of the most important of the intermediaries are the Non-Governmental Agencies or NGOs. Such groups can vary enormously (Figure. 7.2) and be subject to a wide range of loyalties and obligations besides those of the people with whom they work. Often they are accused of facilitating the privatisation of welfare services which ought to be the responsibility of the state, rather than acting as conduits for the participation of the community within the planning and management process. This is said to be particularly true of foreign NGOs, whose priorities may be quite different from the more community based organisations (CBOs) or the looser social movements.

Conclusions

There is no doubt that the pressing problems of the present and the accumulated problems of the past have produced a very complex set

Figure 7.2   A typology og NGOs

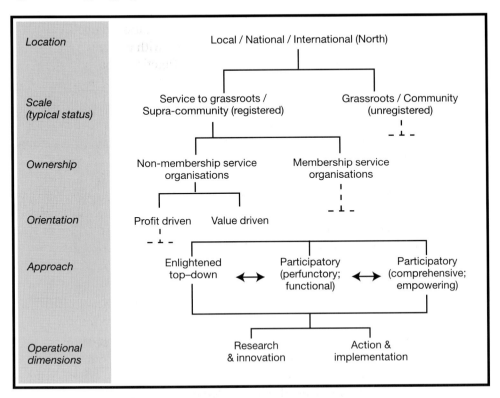

of issues with which urban management has to cope. There is little point in dismissing many of the real problems, such as poverty, inequality, low standards of living, crime and environmental degradation as 'myths' of mega-urbanisation that will eventually disappear in time. Certainly, rapid urbanisation needs to be accepted by national and urban planners as the norm, rather than the exception. Urban managers must recognise and accept the dynamics of the situation and work with rather than against these in order to alleviate the many real problems which exist. The research approaches of academics and practitioners need to reflect this reality (Table 7.4).

Devas and Rakodi (1993) suggest that there are new approaches evident in the planning process – ones which are based on an integration of resource, physical and social management approaches. Planning and management are inherently political processes by which different agencies and groups compete for access to scarce resources, whether these relate to land, finance or service provision. The system for allocating these scarce resources is, therefore, extremely important – as is accountability. While the input from local communities has been

Table 7.4   *Research approaches for sustainable urbanisation*

....

Research philosophy (why research is done)

● research is not value-free and should become positively part of citizen/community empowerment
● research should be based on a dialogue with the community to understand how external knowledge and resources can be co-ordinated with local knowledge and goals for the benefit of the community
● research should enhance local knowledge
● research should show and encourage commitment to democratic practice

Research strategy (how to do research)

● city-based collaborative research: linking exogenous and indigenous participants at national through to local level
● co-operative research bringing findings together across disciplines for discussion and dissemination between nations
● rapid research relying less on assembling detail and using local knowledge through RRA and RUA approaches
● pragmatic research: identifying and addressing specific problems in more detail and establishing lines of accountability for their resolution

Research scales and foci (units of research)

● the household
● inter-household and community organisations
● NGOs (local and non-local) and external agencies
● the state – local and national
● cross-cutting focus interests – gender, class, ethnicity

....

Source: Adapted from Douglass (1922).

recognised as crucial in this process, there is also a clear need to strengthen the role of municipal government in order to integrate the management of locally oriented schemes into a forward-looking and integrated planning process. Table 7.5 summarises the features which may define effectiveness for government at this level.

However, it is not only the efficiency and effectiveness of the planning/management systems which are important, it is also their goals and their responsiveness to the variety of needs within the city. These may be economic, social, political or environmental and are often tackled according to who controls the planning system. Clientelistic forms of service provision need to be replaced by more egalitarian systems,

Table 7.5   *Some features of effective municipal government*

. . . .

- Clear responsibility for the main elements which affect the well-being of the urban population and the efficient functioning of the city: roads and public transport, preventive health care, water supply and sanitation, land-use and development control, and residential area development
- Clear responsibility for constructing, operating and maintaining infrastructure and services
- A clear focus of executive authority and well-defined allocation of administrative responsibilities to the agencies and departments concerned
- access to adequate resources to discharge those responsibilities, notably revenues and skilled staff
- Accountability both in overall political terms and in terms of the design of procedures for carrying out policy formulation and implementation
- Responsiveness, which implies both accountability and room for manoeuvre; the latter requires a capacity for independent action and the exertion of leverage on relevant points in the governmental system

. . . .

Source: After Devas and Rakodi (1993: 282–283).

which will only come about if those without a voice are incorporated into the planning and management process. This is easy to say but more difficult to achieve in a meaningful way which avoids paternalism and further clientelism. Making the planning and management system more responsive and accountable through improved levels of municipal democracy is as essential a part of this process as is participatory planning.

Ultimately the goal of management and planning is sustainable urbanisation, as outlined in Chapter 1 (see also Table 7.6). As UNCHS (1996:422) reminds us, the main criteria for judging the success of urban management are:

- the quality of life of the inhabitants, including their basic needs and human rights
- the scale of non-renewable resource use, including recycling
- the scale and nature of renewable resource use, such as water supplies
- the safe disposal of non-renewable wastes generated by the city

Not all citizens will be satisfied with urban management in this context but the objective should be to maximise this satisfaction throughout the city and to improve it further in the future.

Table 7.6 *The goals of sustainable urbanisation*

....

Meeting the needs of the present . . .

● Economic needs: includes access to an adequate livelihood or productive assets; also economic security when unemployed, ill, disabled or otherwise unable to secure a livelihood.

● Social, cultural and health needs: includes a shelter which is healthy, safe, affordable and secure, within a neighbourhood with provision for piped water, sanitation, drainage, transport, health care, education and child development. Also a home, workplace and living environment protected from environmental hazards, including chemical pollution. Also important are needs related to people's choice and control – including homes and neighbourhoods which they value and where their social and cultural priorities are met. Shelters and services must meet the specific needs of children and of adults responsible for most child-rearing (usually women). Achieving this implies a more equitable distribution of income between nations and within nations.

● Political needs: includes freedom to participate in national and local politics and in decisions regarding management and development of one's home and neighbourhood within a broad framework which ensures respect for civil and political rights and the implementation of environmental legislation.

. . . without compromising the ability of future generations to meet their own needs

● Minimising use or waste of non-renewable resources: includes minimising the consumption of fossil fuels in housing, commerce, industry and transport plus substituting renewable sources where feasible. Also, minimising waste of scarce mineral resources (reduce use, reuse, recycle, reclaim). There are also cultural, historical and natural assets within cities that are irreplaceable and thus non-renewable – for instance, historic districts and parks and natural landscapes which provide space for play, recreation and access to nature.

● Sustainable use of renewable resources: cities drawing on freshwater resources at levels which can be sustained; keeping to a sustainable ecological footprint in terms of land area on which producers and consumers in any city draw for agricultural crops, wood products and biomass fuels.

● Wastes from cities keeping within absorptive capacity of local and global sinks: including renewable sinks (e.g. capacity of river to break down biodegradable wastes) and non-renewable sinks (for persistent chemicals; includes greenhouse gases, stratospheric ozone-depleting chemicals and many pesticides).

....

Source: UNCHS (1996).

Key issues

1. Cities must be managed for the benefit of all of their citizens.
2. International agencies have considerable influence in the management of large cities in developing countries.

3. Effective urban management needs more decentralisation of political and financial power.
4. Urban communities must have more involvement in urban management and planning.
5. Effective urban management must encompass physical and social resource management.

Discussion questions

* Evaluate the assertion that urban management is essentially a political process.
* Urban management in development countries is impossible without international assistance. Is this true?
* Are the better-organised cities usually managed by national rather than local authorities? Why should this be so?
* Is community involvement in urban management too self-serving to be encouraged?
* Do NGOs form an effective management link between urban planners and local communities?

References and further reading

Devas, N. and Rakodi, C. (1993) *Managing Fast Growing Cities*, Longman, London.

Douglass, M. (1992) 'The political economy of urban poverty and environmental management in Asia: access, empowerment and community based alternatives', *Environment and Urbanization*, 4(2): 9–32.

Gibbon, P., Bargura, Y. and Ofstad, A. (1992) *Authoritarianism, Democracy and Adjustment*, Nordic Institute for African Studies, Uppsala.

UNCHS (1996) *An Urbanizing World*, Habitat, OUP, Oxford.

United Nations (1991) *Global Strategy for Shelter to 2000*, UNCHS, Nairobi.

World Bank (1991) *Urban Policy and Economic Development: An Agenda for the 1990s*, World Bank, Washington, DC.

 **Index**

Note: Page numbers followed by an 'm' indicate maps; and those followed by an 'f' indicate figures and tables.